The Regulatory Challenge of Biotechnology

BIOTECHNOLOGY REGULATION SERIES

Series Editor: Han Somsen, *Tilburg Institute of Law, Technology and Society (TILT), Tilburg University and Centre for Environmental Law, University of Amsterdam, The Netherlands*

Biotechnology is a term that provokes a range of differing reactions and views, though most would agree on the necessity of ensuring its careful regulation. The rise and rapid evolution of modern biotechnology has generated significant and diverse regulatory challenges, many of which have significant implications for society as a whole.

This series is designed to comprise work, both collaborative and single-authored, that provides a critical insight into the challenges presented by biotechnology, and the range of regulatory techniques and solutions on offer. The issues confronted in the series will range from agricultural biotechnology, including GMO's, to human genetics, though to intellectual property aspects of biotechnology, such as patents and the TRIPS framework. The European and International dimension of biotechnology regulation will be a constant reference point.

Whilst focussing principally on the legal framework of biotechnology regulation, the series will also draw from related disciplines such as environmental studies, politics and biology, and aims to inform policy as much as to comment on it.

The Regulatory Challenge of Biotechnology
Human Genetics, Food and Patents
Edited by Han Somsen

The Regulatory Challenge of Biotechnology

Human Genetics, Food and Patents

Edited by

Han Somsen

Tilburg Institute for Law, Technology and Society (TILT), Tilburg University and Centre for Environmental Law, University of Amsterdam, The Netherlands

BIOTECHNOLOGY REGULATION SERIES

Edward Elgar

Cheltenham, UK • Northampton, MA, USA

Published by
Edward Elgar Publishing Limited
Glensanda House
Montpellier Parade
Cheltenham
Glos GL50 1UA
UK

Edward Elgar Publishing, Inc.
William Pratt House
9 Dewey Court
Northampton
Massachusetts 01060
USA

A catalogue record for this book
is available from the British Library

Library of Congress Cataloguing-in-Publication Data

The regulatory challenge of biotechnology : human genetics, food, and patents / edited by Han Somsen.
p. cm. – (Biotechnology regulation series)
Includes bibliographical references and index.
1. Biotechnology industries–Law and legislation. 2. Biotechnology industries–State supervision. I. Somsen, Han. II. Series.
[DNLM: 1. Biotechnology–legislation & jurisprudence. 2. Genetic Engineering–legislation & jurisprudence. 3. Genetics, Medical–legislation & jurisprudence. 4. Food, Genetically Modified–standards. 5. Patents–legislation & jurisprudence. QU 33.1 R344 2006]
K3925.B56R444 2006
343′.0786606–dc22 2006011742

ISBN 978 1 84542 489 3

Printed and bound in Great Britain by MPG Books Ltd, Bodmin, Cornwall

Contents

Contributors

Roger Brownsword is Professor of Law at King's College London and Honorary Professor of Law at the University of Sheffield (where he was one of the founding members of the Sheffield Institute of Biotechnological Law and Ethics). He co-authored books with Professor Deryck Beyleveld, including (1986), *Law as a Moral Judgment*, Sweet & Maxwell and (2001), *Human Dignity in Bioethics and Biolaw*, Oxford University Press. He co-edited with Professor Cornish and Dr Llewelyn (1998), *Law and Human Genetics: Regulating a Revolution*, Hart. Supported by a Leverhulme Trust Fellowship, he recently started work on a new book for OUP entitled *Rights, Regulation, and the Technological Revolution*.

Professor Brownsword acted as a specialist adviser to the House of Lords Select Committee on Stem Cells and his recent journal papers in this general area include: (2002), 'Stem Cells, Superman, and the Report of the Select Committee', *Modern Law Review*; (2003), 'Bioethics Today, Bioethics Tomorrow: Stem Cell Research and the "Dignitarian Alliance" ', *University of Notre Dame Journal of Law, Ethics and Public Policy*; (2003), 'An Interest in Human Dignity as the Basis for Genomic Torts', *Washburn Law Journal*; (2004), 'Regulating Human Genetics: Dilemmas of Form and Substance, Doctrine and Design', *Medical Law Review*; and (2004), 'Reproductive Opportunities and Regulatory Challenges', *Modern Law Review*.

In addition to his interest in law, ethics and technology, Professor Brownsword has written extensively about the law of contract and he is co-editor of a four-volume set on *Global Governance and the Quest for Justice*, Hart, 2004.

Justine Burley is Adjunct Associate Professor at the National University of Singapore. She obtained post-graduate degrees in both political philosophy and in neuroscience from the University of Oxford, where she was a Lecturer from 1993–2002. She also held a Simon Fellowship (1998–2002), and was a Fellow of the Institute of Medicine, Law and Bioethics (1996–98), at the University of Manchester. She is the editor of (2004), *Ronald Dworkin and His Critics*, Blackwell; (1999), *The Genetic Revolution and Human Rights*, Oxford University Press; and the co-editor (with John Harris) of (2000), *A Companion to Genetics*, Blackwell and (2004) *A Companion to the*

Ethics of Genetics, Blackwell. Her first monograph, entitled *Genetic Justice*, is forthcoming.

Graham Dutfield is Herchel Smith Senior Research Fellow at Queen Mary Intellectual Property Research Institute, Queen Mary University of London. He was formerly Academic Director of the UNCTAD–ICTSD Capacity-building Project on Intellectual Property Rights and Development. He has published three books: (2003), *Intellectual Property Rights and the Life Science Industries: A Twentieth Century History*, Ashgate; (2002), *Intellectual Property Rights, Trade and Biodiversity: Seeds and Plant Varieties*, Earthscan, and (2004), *Intellectual Property, Traditional Knowledge and Biogenetic Resources*, Earthscan. He has served as consultant or commissioned report author for numerous organizations.

Mary E. Footer was Deputy Director of the Amsterdam Center for International Law (ACIL) until 2005. She is currently Professor of Law at Nottingham University. Her published books include: (2005), *An Institutional and Normative Analysis of the World Trade Organization*, Martinus Nijhoff; with Julio Faundez and Joseph J. Norton (eds) (2000), *Governance, Development and Globalization*, Blackstone Press Ltd; with Joseph J. Norton and Mads Andenas (eds) (1998), *The Changing World of International Law in the Twenty-First Century*, Kluwer Law International. Professor Footer has also published numerous articles and chapters about agricultural biotechnology, including 'Agricultural Biotechnology, Food Security and Human Rights', in Francesco Francioni and Tullio Scovazzi (eds) (forthcoming, 2006), *Biotechnology and International Law*; 'Our Agricultural Heritage: Agricultural Sustainability, Common Heritage and Intergenerational Equity', in Nico Schrijver and Friedl Weiss (eds) (2004), *International Law and Sustainable Development: Principle and Practice*, Martinus Nijhoff.

Neil Gunningham is an interdisciplinary social scientist who specializes in environmental policy and regulatory design. He currently holds Professorial Research appointments in the Regulatory Institutions Network, Research School of Social Sciences, and in the School of Resources, Environment and Society, at the Australian National University. His books include: with Kagan and Thornton (2003), *Shades of Green: Business, Regulation and Environment*, Stanford; with Sinclair (2002), *Leaders and Laggards: Next Generation Environment Regulation*, Greenleaf; and with Grabosky (1998), *Smart Regulation: Designing Environmental Policy*, Oxford University Press. He was previously Foundation Director of the Australian Centre for Environmental Law (1992–2001), Visiting Research Fellow at the London

School of Economics (2001), Visiting and Senior Fulbright Scholar at the Center for the Study of Law and Society, University of California, Berkeley and Research Fellow at the American Bar Foundation, Chicago. He has also been consultant to the OECD on compliance and regulatory design, and to the United Nations Environment Programme on management systems and regulatory reform.

Graeme Laurie is Professor in Law at the University of Edinburgh and Co-director of the Arts and Humanities Research Board (AHRB) Research Centre for Studies in Intellectual Property and Technology Law. His research interests include the role of law in promoting and regulating science, medicine and technology. He has provided advice to, and been consulted by, a number of bodies on matters of technology and law. These include the House of Commons Science and Technology Committee (1995), the Governments of Lesotho (1997) and the Faroe Islands (1999), the Massachusetts State Legislature (1999), the World Health Organization (WHO) (2000), and the Human Genetics Commission (2001). In 2001 he convened a WHO Working Group that produced international guidelines on the establishment and maintenance of genetic databases. He is a member of the Interim Advisory Group on Ethics and Governance for Biobank UK and the Privacy Advisory Committee for Scotland, and serves as an Associate Editor (Law) on the *Journal of Medical Ethics*. His publications include the monograph (2002), *Genetic Privacy: A Challenge to Medico-legal Norms*, Cambridge University Press.

Sara Poli is Research and Teaching Fellow in EC Law and EC Environmental Law, University of Trieste. Her articles include (2004), 'Adoption of International Food Standards within the Codex Alimentarius Commission', *European Law Journal*; (2004), 'The Overhaul of the European Legislation on GMOs, Genetically Modified Food and Feed: Mission Accomplished. What now?', *Maastricht Journal of European and Comparative Law* and with Nicola Notaro (2004), 'Environmental Law', in Piet Eeckhout and Takis Tridimas (eds), *Yearbook of European Law*, Oxford: Clarendon Press.

Colin Scott is Professor of EU Regulation and Governance at University College Dublin. Between 2001 and 2003 he was Senior Fellow in Public Law at the Law Program and Regulatory Institutions Network, Research School of Social Sciences, Australian National University. His analytic interests are in regulation and control, and accountability. He has substantive interests in regulation of government, communications regulation and consumer protection. His books include (co-edited) (2004), *Regulating*

Law, Oxford University Press and (co-authored) (1999), *Regulation Inside Government*, Oxford University Press. He has published widely in international journals and books in socio-legal studies and public administration, among others: (2001), 'Analysing Regulatory Space: Fragmented Resources and Institutional Design', *Public Law*, and (2004), 'Regulation in Age of Governance: The Rise of the Post-Regulatory State', in Jacint Jordana and David Levi-Faur (eds), *The Politics of Regulation*, Edward Elgar.

Paul Street is Lecturer of Law at Nottingham University since 2002, before which he taught in the Law Department at the University of Wales, Aberystwyth and the School of Law at the University of Warwick, from where he also holds a PhD in Law. He writes and researches on a range of issues relating to globalization and the international trading system particularly as they relate to biodiversity; sustainable agriculture; food security; biotechnology; land rights; and international development, focussing predominantly on the impacts that transnational legal developments in these areas pose for the countries and peoples of the 'two-thirds world'. He published a monograph (2005) entitled, *Regulating Global Trade and the Environment*, Routledge, and many articles including (2001), 'Trading in Risk: The Biosafety Protocol, Genetically Modified Organisms and The World Trade Organization', *Environmental Law Review*.

Wolfgang van den Daele, studied law and philosophy in Hamburg, Tuebingen and Munich. He is currently Director of the research unit Standard-setting and Environment at the Social Science Research Center Berlin (WZB), and Professor of Sociology at the Free University of Berlin, Germany. From 1985 to 1987 he was a member of the German Federal Parliament's Select Committee on the Opportunities and Risks of Genetic Engineering. At present, Wolfgang van den Daele is a member of the National Board of Ethics for the Federal Republic of Germany.

Publications in the areas of scientific research, technology assessment and environmental sociology include: with Alfred Pühler and Herbert Sukopp (1997), 'Grüne Gentechnik im Widerstreit', *Agrarwirtschaft: Zeitschrift für Betriebswirtschaft und Marktforschung*; with Friedhelm Neidhardt (eds) (1996), *Kommunikation und Entscheidung WZB-Jahrbuch*, Edition Sigma; and (2001), 'Von moralischer Kommunikation zur Kommunikation über Moral. Reflexive Distanz in diskursiven Verfahren', *Zeitschrift für Soziologie*; with R. Döbert (2004), 'Imaginierte Gemeinschaften. Forderungen und Mechanismen transnationaler Solidarität beim Zugang zu patentgeschützten Medikamenten', in Dieter Gosewinkel, D. Rucht, W. van den

Daele and J. Kocka (eds), *Zivilgesellschaft – National und Transnational, WZB-Jahrbuch*, Edition Sigma.

Bernd van der Meulen is Professor in Law and Governance at Wageningen University (Netherlands), and Director of the European Institute for Food Law. He is President and co-founder of the Dutch Food Law Association. He recently published (2004), *The Right to Adequate Food: Food Law Between the Market and Human Rights*, Elsevier and with Menno van der Velde (2004), *Food Safety in the European Union: an Introduction*, Wageningen University Press. He has published extensively on food law in Dutch journals.

Geertrui Van Overwalle is Professor at the Centre for Intellectual Property Rights (CIR) of the Catholic University of Leuven and the Catholic University of Brussels (Belgium). She teaches Patent Law and Intellectual Property Rights in the Biomedical Sciences in Leuven and 'Plant Breeder's Rights and Biotechnology in Brussels. She has been Visiting Professor at the United Nations University (2000–03), the University of Ruhuna, Matara, Sri Lanka (2000), the Renmin University of China (2001) and Monash University, Melbourne (2003). She was recently appointed a member of the European Commission's Expert Group on Biotechnological Inventions.

She has published numerous articles and monographs relating to patent law and biotechnology including (1997), *The Legal Protection of Biotechnological Inventions in Europe and in the United States. Current Framework and Future Developments*, Leuven University Press and (2002), *Study on the Patenting of Inventions Related to Human Stem Cell Research*, European Communities. More recently she wrote (amongst others) (2005), 'Legal and Ethical Aspects of Bio-patenting: Critical Analysis of the EU Biotechnology Directive', in Peter Drahos (ed.), *Death of a Patent System*, Law Text Publishing, as well as (2005), 'Experience on the Patentability of Human Stem Cells and Stem Cells Derived from Them', in *The Ethics of Patenting Human Genes and Stem Cells*, Copenhagen: Council of Ethics and (2005), 'Protecting and Sharing Biodiversity and Traditional Knowledge: Holder and User Tools', *Ecological Economics*.

Foreword

This book is the first in a series focussing on the regulation of biotechnology. The event of modern biotechnology has given rise to a plethora of regulatory challenges. Whereas modern biotechnology is a relatively new phenomenon, many of these are familiar to regulatory theorists and regulators. Precautionary risk regulation, for instance, was pioneered in European environmental regulation, years before it came to form the backbone of the regulation of GMOs within the EU. Issues of privacy and autonomy that arise in the sphere of human genetics are nothing new to the medical lawyer either.

In all these instances, those faced with responsibilities to regulate biotechnology would do well to investigate if lessons can be learned from previous experience. To be sure, even the search for useful existing models for the regulation of a new technology is a challenging and time-consuming intellectual exercise, as it requires awareness of legal domains that may have little prima facie significance for the problem at hand. For example, researchers struggling with the implications of the emergence of bio-banks for the continued viability of requiring prior bilateral informed consent could possibly find some answers to their questions if they appreciated that prior informed consent, a regulatory communicative instrument, was first developed in medical law as a *bilateral* tool, but later evolved into a *multilateral* instrument in environmental law.

But the intellectual investment needed to identify and then usefully apply existing regulatory instruments to the new regulatory context of modern biotechnology is worth the effort if it means that the open-ended process of designing an effective regulatory framework from scratch can be avoided. This volume contains various chapters that represent attempts to engage in this process of learning from past experience. In doing so, however, we should continuously query *what* previous practice is sufficiently relevant for biotechnology to warrant such an exercise of transposition to new technologies, and who we should entrust with that judgement. This, in good part is a normative question and therefore invites public dialogue. In this respect, scientists tend to be much more inclined to adopt a 'business as usual' approach than wider society. Geneticists often respond with bewilderment to public anxieties about technologies that, in their perception, as a matter of principle are no different from traditional and routinely used practices. Why should we approach the issue of genetic testing fundamentally differently from, say, HIV testing? Should food safety law targeting genetically modified foods be based on an entirely different paradigm, for example, as encapsulated by precaution, rather than that underpinning traditional food law? Are there compelling reasons to maintain that the classical requirements for the patentability of inventions should be re-assessed in the context of the patenting of biotechnological inventions, or that ethical considerations should be reviewed by patent offices? How come that the contested concept of human dignity appears to act as a hurdle for potentially life-saving techniques, whilst at the same time abortion and euthanasia have become widely used medical interventions? These are important and difficult questions, which often mask thorny political and ethical issues.

The fact that different societies answer identical questions in different ways, for example, the US and EU as regards the regulation of genetically modified organisms, also proves that socio-political context matters a great deal in this regard. It is clear that such divides between jurisdictions, but also different paradigms within a single jurisdiction, complicate the regulation of biotechnology. It is precisely because so little consensus exists about regulatory goals, that Gunningham (Chapter 1) argues that regulatory pluralism, which has proved useful in the context environmental regulation, cannot play a similarly useful role for the regulation of biotechnology. Gunningham thus identifies a first important characteristic of biotechnology as a regulatory field: a relative lack of agreed standards or goals that should be pursued. Paradoxically perhaps, this is true. This conclusion can guide regulators in the design of their regulatory approach. More specifically, regulators should in those instances assign a different task to law: law, Gunningham asserts, could become 'procedure oriented rather than directly focused on a prescribed goal, and . . . design self-regulating systems by establishing norms of organization and procedure'.

As pointed out by Gunningham, forms of meta-regulation or reflexive regulation are at first sight attractive for the regulation of biotechnology, as they seem to be able to respond to both the complexity of the problem, as well as to the knowledge asymmetry between government and the biotechnology industry. Indeed, manifestations of meta-regulation can be seen to operate relatively well in those areas of biotechnology policy where public trust is more robust. It is common knowledge that public anxieties, for now at least, focus on agricultural biotechnology more than on human genetics. It is perhaps for this reason that the UK Human Fertilisation and Embryology Authority (HFEA) is effectively carrying out tasks that are embedded in what, in essence, is a system of meta-regulation. However, Gunningham stresses, there remain very serious hurdles to be negotiated before meta-regulation and civil regulation can be of similar value as in environmental policy, in particular because of the lack of public trust that such alternative forms of regulation are likely to be met. Also, biotechnology regulation to an important degree revolves around the regulation of risk and ethical issues, in which the state must play a key role either directly, or indirectly in the meta-regulatory model. Such considerations limit the scope for some of the other models discussed by Gunningham, in particular, civil regulation.

Whereas Gunningham feels that for these reasons the role of the state in biotechnology regulation is likely to remain central, Scott (Chapter 2), on the contrary, infers from the fact that biotechnology policy places so much confidence in the state to exert control over social and economic actors that biotechnology policy has not yet matured into a field of regulation at all. Thus, Scott argues, 'extensive standard-setting by state agencies, risks crowding out the capacities of businesses and civil society organization to participate in or determine the standard-setting processes, and risks missing opportunities to promote "ownership" of regulatory norms'.

Apart from the state, markets, social control and technology itself are modalities of control that, with or without state intervention, do regulate biotechnology. Scott therefore explores effective examples of hybrid forms of control that, in a way, have not altogether different from what regulatory pluralists advocate.

Evidently, outcomes of this exercise will differ depending on the problem and associated social field involved. Just as 'the environment' is much too diverse a policy field to justify meaningful generalizations pertaining to its regulation, the same applies to 'biotechnology'. Like Scott, Brownsword (Chapter 3) seeks to correct over-simplistic notions of regulation. For this purpose, regulation is conceived as having four key dimensions: phasing, pitch, mode and range. Where regulation is first phase its purpose is to control ex ante any given genetic practice. Second phase ex post regulation comes into play when such ambitions have been abandoned and, for example, the issue of compensation arises. The need for third phase regulation arises if second phase regulation is not feasible, etc. Regulatory pitch concerns the way in which regulation seeks to engage with its targets, and which can be moral, practical and behavioural. The different options that can be considered to channel behaviour, which were earlier considered by Scott and which includes a 'technical fix', are what Brownsword terms regulatory modes. Regulation can range from a blanket prohibition to uninhibited permission, or tilt towards either of the two extremes.

In combining these four dimensions, an almost endless spectrum of regulatory options emerges. Two additional corrections are introduced that compensate for crude conceptions of the regulation of human genetics: the focus on 'rogue providers' of genetic services, and the preoccupation with human genetics as a regulatory target. As for the focus on rogue geneticists, Brownsword shows that possibly more serious but certainly more realistic threats stem from perfectly respectable actors, including purchasers of such services, the pharmaceutical industry and insurance companies. Whereas Scott considers techno-regulation a promising regulatory mode in the context of GMO regulation, in the context of human genetics, Brownsword explores 'the nightmare that is techno-regulation'.

In their discussion of appropriate regulatory frameworks to channel developments in human genetics Brownsword and Burley (Chapter 4) both employ a mind experiment as a tool. This approach is widely used when policies must be developed now that will still be appropriate to cope with uncertain future scenarios. This is particularly useful in the context of new technologies, including human genetics (see Punic et al. 2006).*

Burley's is a hypothetical world in which present genetic technologies have advanced significantly, although still within the bounds of what is reasonably possible. In this world, Burley posits that government should be informed by contemporary liberal thought, characterized by its dual emphasis on freedom and equality. The position taken is a prescriptive and normative one, but at the same time one of direct practical application: regulation that falls foul of these two basic values is 'wrong'. This perspective allows Burley to arrive at the conclusion that state-run insurance schemes that mitigate brute luck with respect to individuals' genetic constitutions are to be preferred over private schemes. Using similar reasoning, she argues that it would be illegitimate for regulators to deny citizens access to therapeutic cloning.

The book then turns to the regulation of genetically modified organisms (GMOs) and agricultural biotechnology. Much of the complexity and controversy surrounding the regulation of GMOs concerns the regulation of risk. In the tradition of Beck, central to Chapter 5 by Street is the argument that risk is a social phenomenon. Different constructions and perceptions of risk associated with any given technology thereby exist that may be equally valid. This, in turn, calls into question the legitimacy of current institutional practice, and has triggered institutional strife. Street catalogues the importance of institutional competition in the context of the discussions about the relationship between the Sanitary and Phytosanitary (SPS) Measures Agreement and the Cartagena Protocol on Biosafety. This analysis culminates in a call for more inclusive forms of decision-making about the marketing of GMOs and the risks they pose.

Within the EU, a logical starting point for the operationalization of such ambitions is the precautionary principle. Much, perhaps too much, has been written about this principle that informs the regulation of scientific uncertainty within the EU. However, Van den Daele (Chapter 6) offers a truly challenging and insightful analysis of principle. His starting point is that what really is at issue is not so much the regulation of risk, but the democratic control over the forces of social change associated with biotechnology. Escalating risk perceptions and the precautionary principle have come to serve as a strategy for states to gain political control over technology within a liberal regime of innovation. Closer analysis of the precautionary principle suggests that the expectations of critics who believe that it will foster increased democratic control and imply a ban on GMO products may be misplaced.

Central to Van den Daele's chapter, however, is the thesis that its application in the sphere of GMOs shows that the precautionary principle is no longer a *legal* principle. Instead, evidenced by a number of notorious cases, the precautionary principle has become politicized in a way to suppress innovation that is deemed undesirable for reasons that have nothing to do with risk. Political planning is thereby taking place under the guise of risk precaution. The Commission and the European Court of Justice (ECJ) appear to have understood that this position is untenable, however, and have responded in a way to restore the distinction between politics, the exclusive preserve of parliaments, and risk regulation carried out by regulatory agencies. In any event, Van den Daele argues, the moratorium has been insignificant as a measure of precaution, and any impression that it symbolizes the sovereignty of politics over science is false. Neither should we assume that the moratorium has been in the public interest. At the very least, it represents considerable wasted political capital that could have been usefully applied to address more pressing and proven agricultural problems.

The current EU regulatory regime for GMOs for human consumption discussed by Van den Daele is assessed in more detail by Van der Meulen (Chapter 7). Judged by criteria such as legal certainty for applicants and the public, transparency of the authorization process, and the proportionality of, in particular, provisions on traceability, labelling, coexistence and segregation, his conclusion that this regime leaves much to be desired is a persuasive one.

Sara Poli (Chapter 8) focusses her discussion on national coexistence policies and the numerous issues of EC law to which they give rise. Measures intended to create GMO-free zones such as those adopted by a number of regions in Austria, and which are the product of strong and almost universal local opposition to GMOs, must pass the test of Article 95(5) EC. In the light of earlier cases involving the adoption of more stringent national rules after EC harmonization, the sorry fate of the Austrian rules was perhaps predictable. The Austrian case appears to indicate that the Commission is no more sympathetic towards protective national rules concerning GMOs than other national initiatives that depart from the internal market paradigm.

Nonetheless, in the context of coexistence the Community only claims competences over aspects that concern issues related to the environment and health. This has given rise to (failed) national initiatives based on Article 23 of Directive 2001/18/EC on the Deliberate Release into the Environment of Genetically Modified Organisms (Deliberate Release Directive OJ L106/1 2001), which allows Member States to take appropriate measures to avoid the unintended presence of GMOs in other products. Economic aspects explicitly fall within Member States' reserved powers, but in practice this may prove neither a clear nor sufficient basis for Member States to adopt appropriate coexistence measures. All in all, there is every reason to expect future political conflict over national coexistence policies.

Mary Footer (Chapter 9) addresses a phenomenon associated with biotechnology that is of global concern: the commodification or 'enclosure' of the common heritage of mankind that is represented by, on the one hand, crop germplasm and, on the other hand, by what may be termed socio-agricultural pluralism. Her historical and legal analysis lays bare two distinct processes of enclosure that operate in respect of these different commons. Driving forces are, respectively, claims of permanent sovereignty and ownership over natural resources and mechanization of agriculture, and multilateral trade instruments including Trade-related Aspects of Intellectual Property Rights (TRIPS). Both types of commons are in danger of disappearing permanently if these processes are to follow their natural course. This leads us to the role of patent law in agricultural biotechnology. Brownsword and Scott, in different contexts have each offered different perspectives on techno-regulation. Dutfield (Chapter 10) explores a specific and (economically) significant example of techno-regulation, so-called 'terminator technology', a technological fix that protects and enforces intellectual property rights (IPR) over patented crops. Three questions arise in connection with this technology: how will it affect the future of agriculture, should we encourage it and, if not, are there sufficient grounds to conclude that no patents should be granted to protect it?

As for its potential impact, Dutfield concludes that 'terminator technology has the genuine potentional to seriously disrupt poor world agricultural systems that support the livelihoods of hundreds of millions of people'. As for the question whether this kind of technology should be patentable, however, attention is drawn to the paradox that a negative answer might actually encourage research in this area, since it is especially in jurisdictions where IPR protection is weak that the technology would be useful. In any event, there are other, perhaps more appropriate ways to regulate terminator technology, most notably through laws forbidding it. Once again, this chapter concludes with a warning about the socio-economic implications of terminator technology, especially for developing country farmers. Similarly, patents have radically altered the landscape of medical research involving genetics. A common fear is that patents granted, for example, to Myriad for its BRCA1 and BRCA2 genes have stifled much-needed medical research into breast and ovarian cancer. In Chapter 11, Laurie concentrates on the question whether research subjects must consent to the patenting of research in which they have participated as a precondition for the granting of such a patent.

The question first gives rise to a critical evaluation of the importance of informed consent in general, and in a research context in particular. In this respect it is noted that the fundamental issue at stake is not the obtaining of consent, but the furnishing of respect. In as far as there are alternative ways of showing respect, consent may not always be required. It is also important to realize that rules on consent have been developed in the context of medical care and treatment, and that in a research context different standards of disclosure should apply. These observations imply that a requirement of prior informed consent for patents emanating from medical research involving humans is not self-evident. The ECJ (European Court of Justice), in Case 377/98, *Netherlands* v. *Parliament and Council* [2001] ECR I-7079 has ruled so as to shield the patent system enshrined in Directive 98/44/EC on the Legal Protection of Biotechnological Inventions (Biotechnology Directive) against the introduction of an additional condition in the shape of prior informed consent. The purpose served by such an additional condition is already sufficiently protected by regulation.

However, the Court's ruling does not settle the issue. In part this is because a narrowly drafted consent requirement in patent law, that is, one that strictly concerns the filing of a patent, would avoid usurpation of the regulatory system the ECJ was so eager to avoid. Also, as the infamous *Moore* case illustrates, although civil remedies may compensate aggrieved research subjects, this leaves the validity of the patent unaffected. Laurie argues that, since patents are about incentives, there is a case for introducing a threat that patents may be invalidated on the grounds that they were obtained without prior informed consent. Moreover, based on the premise that there should be congruence between the patent system and regulation, there is much to be said for incorporating the vehicle of consent in patent law as an expression of respect for human dignity and integrity that governs both systems.

All these considerations ultimately lead Laurie to argue for a limited role of prior informed consent in patent law. The fact that, in practice, those who wish to participate in research often will have no choice but to accept commercialization of the outcomes of the research, does not affect the importance of that conclusion.

Concluding the book is Chapter 12 by Van Overwalle, who offers a panoramic view of objections against the patenting of biological material. Such objections focus on the subject matter, which is not the result of a creative process and could be ethically objectionable, and non-compliance with the substantive patentability requirements of novelty, inventive step and industrial applicability. Reservations against the patentability of biological material also relate to the impossibility to comply with the requirement of disclosure: often it proves impossible to describe the process of making the end product, or to repeat the original process of making. Concerns about patents for biotechnological inventions also concern their impacts, in particular on the freedom of research but also on agricultural practice and the dissemination of technology more generally.

Having surveyed public anxiety about biotechnology patents, Van Overwalle does not take the position so often adopted by specialists in patent law that these concerns can be best addressed through the regulatory process. Instead, she explores tools that patent law offers that may alleviate these concerns. These include use of the possibility offered by Article 27.3 of TRIPS that allows for the exclusion of plants and animals from patentability, the exclusion of genetic methods of testing, the granting of patents on end products only, reconsideration of morality clauses such as Article 6 of the Biotechnology Directive, and the more radical idea of developing a compensatory liability regime.

Alternatively or simultaneously, patent offices could apply more stringent patent standards. Patent offices could also reduce the scope of patent claims that are excessively broad. The idea of introducing additional patent requirements such as informed consent was discussed by Laurie. The Biotechnology Directive offers a legal basis for this approach, which could also justify introduction of an origin requirement.

Detrimental effects of biotechnological patents on research and health care may be prevented or mitigated through research exemptions, which in Europe are part of patent law, but are applied differently from country to country. Similarly, the denial of patents for methods for diagnostic treatment alleviates negative effects of health care patenting. Compulsory licensing to address health concerns is an option that is gaining interest and finds its legal basis in Article 31 of TRIPS.

All these options are located within the system of patent law itself. However, from the perspective of public participation and transparency, it is attractive to consider possible mechanisms outside the patent system. Indeed, voluntary codes of conduct or other private standards are increasingly widely used, and offer prospects of both reinforcing and supplementing the current patent regime.

Such public participation in genetic governance is increasingly regarded as a discrete value in democratic society. It is also acknowledged that the quality and effectiveness of regulation is served by deliberative models of governance. Without exception, these chapters acknowledge and, in different modalities, advocate public participation in the resolution of the countless political, social and ethical issues to which modern biotechnology gives rise. This volume may be seen as a contribution to that endeavour.

Han Somsen
Series Editor

NOTE

* Punic, Y., S. Delaitre, I. Maghiros and D. Wright (eds) (2006), *Deliverable D2 – Dark Scenarios in Ambient Intelligence: Highlighting Risks and Vulnerabilities*, Safeguards in a World of Ambient Intelligence (SWAMI) Project.

Acknowledgements

Many individuals and academic institutions have played an important role in the realization of this volume. The conference where these chapters were first discussed was organized by Yvonne ter Horst, research assistant at the Research Unit for Biotechnology & Regulation Amsterdam (RUBRA), based at the Centre for Environmental Law, University of Amsterdam. Yvonne also carried out much of the editing work that was involved in the production of this book. I also wish to express my gratitude to Rose Dammen for her assistance. Annemarie Sprokkereef lent practical assistance during the conference, and finalized the chapters for publication. Stefan Dimitrov, researcher at RUBRA, has remained patient and loyal during all these hectic months, and provided practical and moral support where needed.

RUBRA has been established to carry out a multi-annual interdisciplinary research project generously financed by the Netherlands Organisation for Scientific Research (NWO), of which the present volume forms part.

The European University Institute in Florence granted a Jean Monnet fellowship that assisted the completion of the book. Special thanks are owed to Francesco Francioni, Christian Joerges and Neil Walker for their time and general support.

I also wish to thank my colleagues Bert-Jaap Koops and Corien Prins of Tilburg Institute for Law, Technology and Society (TILT), for their enthusiasm and help during my first months there. Finally and most importantly, I am very grateful to the authors for producing these stimulating and thought-provoking chapters.

Han Somsen
Series Editor

Abbreviations

ABI	Association of British Insurers
ACIPA	Australian Centre for Intellectual Property in Agriculture
ACRE	Advisory Committee on Releases to the Environment
ACTG	Advisory Committee on Genetic Testing
AIA	advanced informed agreement
ANU	Australian National University
CAP	Common Agricultural Policy
CBD	Convention on Biological Diversity
CCIPR	Collective Community Intellectual Property Right
CCTV	closed circuit television
CFI	Court of First Instance
CGIAR	Consultative Group on International Agricultural Research
CGRFA	Commission on Genetic Resources for Food and Agriculture
CORE	Comment on Reproductive Ethics
COW	crude oil washing
DSB	Dispute Settlement Body
EC	European Community
ECtHR	European Court of Human Rights
EFSA	European Food Safety Authority
EG	embryonic germ stem cell
EJC	European Court of Justice
EPC	European Patent Convention
EPO	European Patent Organisation
EPO	European Patent Office
ES	embryonic stem
ESRC	Economic and Social Research Council
ESHG	European Society of Human Genetics
EU	European Union
FAO	Food and Agriculture Organization
FSC	Food Stewardship Council
GAIC	Genetics and Insurance Committee
GATT	General Agreement on Tariffs and Trade
GM	genetically modified

GMCs	genetically modified crops
GMOs	genetically modified organisms (or micro-organisms)
GURTs	genetic use restriction technologies
HFEA	Human Fertilisation and Embryology Authority
HLA	human leukocyte antigen
IARCs	International Agricultural Research Centres
ICTSD	International Centre for Trade and Sustainable Development
INPO	Institute of Nuclear Power Operations
IPR	intellectual property rights
ITPGRFA	International Treaty on Plant Genetic Resources for Food and Agriculture
IVF	in vitro fertilization
LMO	living modified organism
MDM	market-driven manufacturing
MEA	multilateral environment agreement
MTA	Material Transfer Agreement
MTN	Multilateral Trade Negotiations
NARCs	National Agricultural Research Centres
NGO	non-governmental organization
NIH	National Institutes of Health (US)
NSMD	non-state market-driven
OILPOL	International Convention for the Prevention of Pollution of the Sea by Oil
PBRs	Plant Breeders' Rights
PGD	preimplantation genetic diagnosis
PGR	plant genetic resources
PGRFA	plant genetic resources for food and agriculture
PIAG	Patient Information Advisory Group
PP	precautionary principle
RAFI	Rural Advancement Foundation International
SARD	sustainable agriculture and rural development
SBT	segregated ballast tank
SCNT	somatic cell nuclear transfer
SPS	Sanitary and Phytosanitary Measures (WTO Agreement)
TBT	Technical Barriers to Trade Agreement
TIP Right	Traditional Intellectual Property Right
TRIPS	Trade-related Aspects of Intellectual Property Rights
UNCTAD	United Nations Conference on Trade and Development
UNDP	UN Development Programme
UPOV	Union Internationale pour la Protection des Obtentions

	Végétales (International Union for the Protection of New Varieties of Plants)
USDA	United States Department of Agriculture
WTO	World Trade Organization
WWF	World Wide Fund for Nature, formerly known as World Wildlife Fund
ZKBS	Central Commission for Biological Safety (Zentrale Kommission für die Biologische Sicherheit)

PART I

General perspectives on biotechnology regulation

1. Regulating biotechnology: lessons from environmental policy

Neil Gunningham

In developing principles for the regulation of biotechnology, what can we learn from environmental policy? In particular, are the frameworks, perspectives and insights that have enriched environmental regulation also helpful in curbing the food safety and environmental risks posed by biotechnology?[1]

This chapter examines four different frameworks, each of which offers very different policy prescriptions and asks to what extent each of them might be useful in the context of regulating biotechnology. It does so in three parts. First, it locates the various frameworks within the context of wide-reaching political and economic changes of the last two decades. Second, it examines smart regulation, meta-regulation, civil regulation and the 'license perspective', and the insights that each might provide to the regulation of biotechnology as it relates to food safety and environmental protection. Finally, and briefly, it draws some broader but provisional conclusions.

1 RECONFIGURING REGULATION: A CONTEXT

The 1980s and 90s saw a resurgence of free-market ideology, which, assisted by the economic and political collapse of the former Soviet Union, enabled neo-liberalism to triumph almost unchallenged for most of that period and beyond. And while public opposition precluded the sort of wholesale deregulation that occurred in some other areas of social regulation, environmental regulatory budgets were substantially cut in almost all jurisdictions. This trend shows little sign of changing under the lower taxation regimes that now characterize the large majority of economically advanced states, irrespective of the party in power.

But while government regulators have been losing both their power and resources, others have begun to fill the regulatory space they previously occupied. For example, environmental non-governmental organizations

(NGOs), aided by advanced techniques for information-gathering (from digital cameras to satellite imaging), have become increasingly sophisticated at communicating their message (via global television, international newspapers and the internet) and in using the media (and sometimes the courts) to amplify the impact of their direct action campaigns. They have not only sought to shape public opinion, to lobby governments and to pressure industry directly, but also to influence consumers and markets through strategies such as orchestrating consumer boycotts or preferences for green products.

At the same time, a variety of commercial third parties have also begun to take a considerable interest in environmental issues. Banks and insurance companies seek to minimize their financial risk by scrutinizing more closely the environmental credentials of their clients. And financial markets themselves have become responsive to good or bad environmental news, rewarding environmental leaders with a share price increase and discounting the share price of laggards. So too is supply chain pressure increasingly important.

As part of this reshaping of the regulatory landscape, a number of environmental stakeholders have to some extent departed from their traditional roles. Some business groups, such as the World Business Council for Sustainable Development, have become proactive, arguing that business is part of the solution rather than merely the problem, and sought to develop a variety of voluntary initiatives through which business seeks to shape its own environmental destiny. Environmental NGOs, frustrated with their limited impact on governments, or at the ineffectiveness of government in protecting the environment, have redirected their attention towards corporations through strategies ranging from confrontation to partnership.

What has evolved is not a retreat of the regulatory state and a return to free markets but rather a regulatory reconfiguration. Negotiated agreements in Western Europe, a plethora of informational regulation initiatives, various forms of industry self-management and a variety of enterprises (commonly using supply chain and financial market pressure) built around harnessing third parties as surrogate regulators, nevertheless involve a continuing government role. Even in relation to problems that the state is ill equipped to address directly it almost invariably retains a supporting role, underpinning alternative solutions and providing a backdrop without which other, more flexible options, would lack credibility, and stepping in where they fail. That is, in almost all circumstances the state is still involved in engineering solutions to environmental problems rather than trusting the market, unaided, to provide them. The following sections provide a broader perspective on this regulatory reconfiguration.

2 REGULATING BIOTECHNOLOGY: FOUR FRAMEWORKS

The following section examines four different frameworks, or lenses, through which one might better understand environmental policy and regulatory design. None of these lenses offers (or necessarily purports to offer) a complete prescription for what the next generation of policy instruments should involve. However, as will be shown, individually and collectively, they enrich our understanding of the options available to decision-makers, and some of these frameworks in particular, provide valuable insights concerning the challenges facing biotechnology regulation and how they might best be resolved.

However, the 'fit' between these frameworks, designed to meet environmental goals, and the needs of biotechnology regulation, is far from complete. As will be apparent, the political, social, economic and scientific contexts in which biotechnology and environmental regulation must operate are very different. For biotechnology, regulation must take account of high scientific uncertainty, high risks, a substantial imbalance of knowledge and power between the private sector and the state, diametrically opposed interests (and absolutely no consensus on either preferred outcomes or how best to achieve them), and in Europe at least, a breakdown of public trust.[2]

2.1 Smart Regulation and Regulatory Pluralism

Traditionally, regulation was thought of as a bipartite process involving government and business, with the former acting in the role of regulator and the latter as regulatee. However, a substantial body of empirical research reveals that there are a plurality of regulatory forms, that numerous actors influence the behavior of regulated groups in a variety of complex and subtle ways (Gunningham and Grabosky 1998; Rees 1988, p. 7), and that mechanisms of informal social control often prove more important than formal ones (Gunningham 1991).

These insights have led some policy-makers to investigate how public agencies may harness institutions and resources residing *outside* the public sector to further policy objectives in specific concrete situations. This approach can be seen as part of the broader transition in the role of governments internationally: from 'rowing the boat to steering it' (Osborne and Gaebler 1992), or choosing to 'regulate at a distance' by acting as facilitators of self- and co-regulation (Grabosky 1995), rather than regulating directly. Thus for regulatory pluralists, environmental policy-making involves government harnessing the capacities of markets, civil society and other institutions to accomplish its policy goals more effectively, with greater social acceptance, and at less cost to the state (Gunningham et al. 1999). And since parties and instruments interact with each other and with state regulation in a variety of ways, careful regulatory design ('smart regulation') will be necessary to ensure that pluralistic policy instruments are mutually reinforcing, rather than being duplicative, or worse, conflicting (Gunningham and Grabosky 1998, Chapter 6).

But while this framework has provided some valuable insights that have been applied to considerable effect in the area of environmental regulation,[3] its application to biotechnology regulation is problematic. Indeed, on close examination, a number of the design principles of smart regulation (Gunningham and Grabosky 1998, Chapter 6) have only very limited application when it comes to biotechnology, and the range of credible policy options in the latter domain is relatively narrow. For example, the principles of smart regulation suggest the use of a broad range of policy instruments, which might include information and education, voluntarism, self- and co-regulation, market-based instruments and direct regulation. But many of these instruments manifestly have little or no role to play in biodiversity regulation: no amount of information or education will influence regulatees (biotechnology companies), nor, given the gap between the self-interest of those companies in rapid commercialization and the public interest in minimizing the unanticipated consequences of biotechnology, is self-regulation a credible option.[4] And market-based instruments are largely incapable of providing the sorts of signals necessary to change their behavior (although they might influence consumer behavior). This leaves direct government regulation (primarily through a system that includes mandatory pre-market risk assessment and approval, and the establishment and monitoring of safety standards), liability rules (capable of sending powerful economic signals about the consequence of failure to meet the legal standard) and perhaps informational regulation (product labeling) and co-regulation in limited areas – but not much else.[5]

Smart regulation also advocates the involvement of a broader range of actors than just government (as regulator) and industry (as regulatee). But again the circumstances of biotechnology constrain the extent that this is practicable. Certainly there is a potential role for public participation in parts of the decision-making process, for greater transparency and for the use of outside scientists to provide expert scientific advice. But there is little scope for harnessing surrogate regulators such as financial institutions and markets, or industry associations, and while there is a crucial role for NGOs (explored under civil regulation below) this is essentially as a countervailing force – indeed if the state confined itself to steering not rowing, then fundamental pillars of effective biotechnology regulation (risk assessment, mandatory pre-market approval, established safety standards etc.) identified elsewhere, would be undermined.

Another recommendation of smart regulation is instrument sequencing: introducing less interventionist measures first, but with a clear message that they will be reinforced by more draconian measures should they be unsuccessful. But while this strategy can work well with issues such as pollution control and ozone-depleting substances (where there is latitude for experimentation without disastrous consequences should initial efforts fail) this is not the case for biotechnology, where policy failure may lead to irreversible loss or catastrophic consequences. Once the biotechnology genie is out of the bottle (or the genetically modified [GM] crop has escaped into adjoining fields) it is very difficult to get it back, and the consequences if the initial policy instrument fails are unacceptable.

Whether smart regulators can capitalize on opportunities for win-win solutions is again problematic. While the biotechnology companies would argue that GM products are not materially different from any other products, and that they offer a cost-effective means of feeding the poor and solving world poverty, many other groups disagree with almost everything the biotechnology companies are trying to do. With biotechnology it is not a case of agreement on the ends, with opportunities to explore least-cost or win-win means of achieving them, but rather of fundamental (often ethically based) disagreements about ends themselves.

Certainly some smart regulation design principles do resonate strongly with biotechnology regulation. For example, there is considerable scope for empowering forces of civil society to act as surrogate regulators (as we will see when examining civil regulation below), and there is also scope for informational regulation in the guise of product labeling. But overall, the design principles articulated in smart regulation have only limited insights to offer when it comes to biotechnology regulation.

Finally, there is a broader danger in seeking to apply smart regulation to the area of biotechnology. Smart regulation is a normative approach. Its vantage point is 'good' public policy. It says in effect: 'tell me what policy objective you *want* to achieve and here are some principles that will help deliver on those goals – effectively, efficiently and in ways that are politically acceptable' (Gunningham and Grabosky 1998, pp. 22–4). But it is only when there is some degree of agreement on preferred policy objectives, that it becomes practicable to apply this perspective, and to explore through smart regulation the best means to achieve these objectives.

Yet when we turn to biotechnology, the policy agenda is shaped by a variety of political agendas and there is very little agreement on policy objectives. In many (but by no means all) areas of environmental protection (for example, pollution control, and natural resource management, but not climate change) the debate today is largely about how to achieve broadly agreed objectives (can we get reasonably clean air and water by relying on market mechanisms, or voluntary agreements, or do we need regulation).

But with biotechnology, the protagonists are diametrically opposed on fundamental issues. For example, some 'moderates' would argue that the policy goal for biotechnology, as it relates to environment and food safety issues, is (1) a regulatory system that ensures safe products (to humans and the environment) and engenders public trust (2) without unnecessary burdens that prevent the benefits of transgenic crops being realized. But many NGOs would dispute the second of these objectives, and in the US at least, industry would likely agree in principle with the first but argue that in applying it, GM crops should be treated no differently from any others. Moreover, in such a politically charged arena, policy issues get squeezed to the periphery. Compare the European Union (EU) and United States approaches to biotechnology regulation, which are dramatically divergent and explicable largely in terms of the power of different political constituencies, and the influence of different institutional environments, rather than in terms of 'rational' policy choices to achieve an agreed objective.

2.2 Reflexive and Meta-regulation

The literature on reflexive law[6] recognizes that the capacity of the regulatory state to deal with increasingly complex social issues has declined dramatically. As Teubner (1983) and others (Teubner et al. 1994) have argued, there is a limit to the extent to which it is possible to add more and more specific prescriptions without this resulting in counterproductive regulatory overload. Traditional command and control regulation (a form of 'material law')[7] is seen as unresponsive to the demands of the enterprise and unable to generate sufficient knowledge to function efficiently. In sum: the complexity of society outgrows the possibilities of the legal system to shape the complexity into a form fitting to the goal-seeking direct use of law (Koch and Nielsen 1996). To give a concrete example, one cause of the Three Mile Island nuclear accident and near meltdown, was that operators simply followed rules, without any capacity for strategic thinking, and as events unfolded that were not covered by a rule, they had no capacity to read the situation and respond appropriately.[8]

In contrast, reflexive[9] regulation, which uses *indirect* means to achieve broad social goals, has, according to its proponents, a much greater capacity to come to terms with increasingly complex social arrangements. This is because it:

> focuses on enhancing the self-referential capacities of social systems and institutions outside the legal system, rather than direct intervention of the legal system itself through its agencies, highly detailed statutes, or delegation of great powers to the courts . . . [it] aims to establish self-reflective processes within businesses to encourage creative, critical, and continual thinking about how to minimize . . . harms and maximize . . . benefits. (Orts 1995, p. 1232)

Put differently, reflexive regulation is procedure-oriented rather than directly focused on a prescribed goal, and seeks to design self-regulating social systems by establishing norms of organization and procedure (Fiorino 1999).

Such a strategy can also be viewed as a form of 'meta-risk management' whereby government, rather than regulating directly, risk manages the risk management of individual enterprises.[10] This is what happens under the 'safety case' regime, instituted on North Sea oil rigs following the Cullen enquiry into the Piper Alpha disaster where 167 lives were lost (Cullen 1990). This involves what is in effect a safety and risk management system being developed by the rig operator and submitted to the regulator for scrutiny and approval. Similarly, the safety regime established for the nuclear power industry, post-Three Mile Island, ceased to be primarily about government inspectors checking compliance with rules, and more about encouraging the industry to put in place safety management systems that were then scrutinized by regulators, and in this case, by the industry association in the form of the Institute of Nuclear Power Operations (INPO) (Braithwaite and Drahos 2000; Rees 1994). Under this approach, which is most developed in Christine Parker's *The Open Corporation*: 'the role of legal and regulatory strategies is to add the "triple loop" that forces companies to evaluate and report on their own self-regulation strategies so that regulatory agencies can determine that the ultimate objectives of regulation are being met' (Parker 2002, p. 255). Such a government role is crucial because while companies may have the potential for effective self-regulation they do not necessarily have either the incentive to engage in this approach nor the systems in place to ensure that it is effective (as the StarLink corn fiasco clearly demonstrates; see Bratspies 2003).

This approach has a considerable attraction when it comes to regulating biotechnology. First, biotechnology is a complex problem that does not lend itself to prescriptive regulation as Newell and Glover (2003, p. 8) point out:

> Regulatory systems are slow to evolve and modify, and governments often react to technological change in the private sector rather than drive it. The practical difficulties of tracking cross-border trade in GMOs [genetically modified organisms] monitoring where such crops are being grown, and enforcing biosafety

> regulations at farm level, presents technical, logistical and administrative challenges to even the most developed countries.

Second, there is a substantial imbalance of knowledge between government and the biotechnology industry, with the industry being far more capable of identifying the risks – and arguably of managing them – than the regulators. However, given the industry's self-interest in early commercialization, it is only if it is given an incentive to risk manage, and only if its risk management is closely scrutinized by government (and by independent scientific experts, and with considerable transparency) with the threat of more direct intervention if it fails, that it is likely to take effective action.

On this model, what is needed is the risk management of risk management, or put differently, independent risk assessment on the basis of information supplied by the companies. Arguably, there is little choice but to go down this path, given the imbalance of knowledge and resources between the companies and the state. As Newell and Glover (2003, p. 3) put it:

> Biotechnology firms are, in many ways, the 'street level bureaucrats' of biotechnology, those expected to enforce and implement government regulations regarding biotechnology products. Not only are they the front-line producers and distributors of the technology, a fact which places them well to provide insights and channel their experience into the design of regulatory systems, but the in-house scientific expertise they have and the level of capital they own, make them key advisers and powerful political players . . .

In such circumstances, Parker in particular would emphasize the importance of connecting the internal capacity for corporate self-regulation with the internal commitment to self-regulate, something that is 'done by inducing a corporate crisis of conscience through regulatory enforcement action, legal liability and public access to information about corporate social and legal responsibility' (Parker 2002, p. 246). Having achieved this result, Parker argues that regulators should hold corporate self-regulation accountable and facilitate the potential for other institutions of civil society to hold it accountable, 'by connecting the private justice of internal management systems to the public justice of legal accountability, regulatory coordination and action, public debate and dialogue, . . . through providing self-regulation standards against which law can judge responsibility, companies can report and stakeholders can debate' (Parker 2002, p. 276). This implies that corporate structures be permeable, and open to deliberations with a broad range of stakeholders.

However, when it comes to biotechnology, the gap between the metaregulation ideal, as Parker conceives it, and the reality, may be so large as to make this approach more attractive on paper than in practice. Baldwin

(2004) in particular, points to some obstacles to the effectiveness of meta-regulation, which may be far more serious in some circumstances than in others. First, it may be very difficult to persuade corporate managers to view the world in anything approximating the same way as regulators. For example, they may view non-compliance not as a failure on their part but as an issue of risk management, to be responded to in a way that maximizes shareholder returns. As such it may be something to be managed as part of protecting the enterprise's social license or competitive position. Thus in the case of biotechnology, enterprises may prefer to invest in public relations campaigns to minimize reputation losses caused by regulatory sanctions, or to focus on concealing activities liable to give rise to regulatory sanctions risks, rather than to ensure compliance in circumstances where it would be economically unattractive to do so.

Baldwin also raises the possibility that it is not possible to 'stimulate corporate self-regulation in a manner that produces coherence and harmony between corporate and social ends rather than confusion and conflict' (Baldwin 2004, pp. 378–9), and that the sort of permeability that Parker views as crucial may not be practicable given, in the case of biotechnology, the massive gap in values and world views, between biotechnology companies and NGOs. There are also questions whether permeability and deliberation will lead to 'regimes of high-cost, high-friction management that are characterised by delays, obfuscations, fudges, indecisiveness, confusion and inaction' (Baldwin 2004, p. 379), and whether the corporation will see any attractions in the deliberative model or simply resist its imposition.

Finally, even if meta-regulation was indeed a viable regulatory option, it remains an open question whether it would be perceived to be credible or merely as another variant of self-regulation vulnerable to tokenism, and whether such a regulatory regime would regain public trust in government oversight of biotechnology safety, which in Europe is at a low ebb.

2.3 Civil Regulation

As defined by Murphy and Bendell (1998, p. 8): 'civil regulation is where organizations of civil society[11] such as NGOs set the standards for business behaviour. Enterprises then choose to adopt or not to adopt those standards'. Those who advocate a greater role for civil regulation argue that the regulatory state is starved of resources, lacking in political will, and incapable of reaching the many businesses that can now operate outside national territorial boundaries. The goal of civil regulation[12] is to fill the vacuum left by the contracting state and to compensate for 'the deficit of democratic governance that we face as a result of economic globalization'

(Bendell 2000, p. 201). As such, there is considerable overlap between this perspective and some aspects of regulatory pluralism, discussed above.

Under civil regulation, the various manifestations of civil society act in a variety of ways to influence corporations, consumers and markets, often bypassing the state and rejecting political lobbying in favor of what they believe to be far more effective strategies. Sometimes NGOs take direct action, usually targeted at large reputation-sensitive companies. Greenpeace's campaign against Shell's attempted deep sea disposal of the Brent Spar oil rig, is one example. Sometimes, they boycott products or producers deemed to be environmentally harmful, as with the effective boycott of Norwegian fish products organized by Greenpeace in protest against that nation's resumption of whaling. Market campaigning, focusing on highly visible branded retailers, is a particularly favored strategy.[13] Less so, are campaigns that seek to provide a market premium for 'environmentally preferred' produce, due largely to the unwillingness of consumers to support such a strategy. More recently, certification programmes such as the Forest Stewardship Council (FSC) are 'transforming traditional power relationships in the global arena. Linking together diverse and often antagonistic actors from the local, national and international levels . . . to govern firm behavior in a global space that has eluded the control of states and international organizations' (Gereffi et al. 2001, pp. 64–5). See also Gamble and Ku (2000).

However, the evolving role of civil regulation has not taken place entirely divorced from state intervention. On the contrary, either in response to pressure from the institutions of civil society or in recognition of the limits of state regulation, governments are gradually providing greater roles for communities, environmental NGOs and the public more generally. Thus a number of next generation policy instruments are geared to empower various institutions of civil society to play a more effective role in shaping business behavior. In effect, they facilitate civil regulation (and regulatory pluralism). These initiatives include public participation provisions under the various US Reinventing Environmental Regulation initiatives, Community Right to Know legislation, some voluntary agreements that contemplate a significant role for third parties, and some forms of environmental partnership in which the public, or public interest groups, are major players.

In the context of biotechnology, globalization has empowered NGOs to act as an important countervailing force. Many NGOs believe that governments have been overly influenced by the multinational biotechnology companies, and that there is a need for global acceptance of norms such as the precautionary principle, transparency and sustainability. The strategies by which they seek to change the behavior either of the biotechnology companies directly, or the way governments police the industry, are many and various. They include direct action; information campaigns (distributing negative information about firms' business practices); consumer boycotts; political discourses (linking the norms of precaution, consumer sovereignty and transparency, to demands for segregation and labeling of GM food); lobbying key retailers (for example, pressuring McDonald's restaurants to reduce demand for GM food); influencing inter-government forums (to establish stricter trade regulations); as watchdogs/monitoring, and by litigation (see Prakash and Kollman 2003, p. 637).

But civil regulation is far from being fully effective in a 'standalone' role, and it acts primarily as a countervailing force: it can lobby, impose boycotts, take remedial action after the event, but it cannot perform key gatekeeping roles. These include mandatory pre-market approval, establishing safety standards, independent risk assessment, and enforcement etc. Such functions can only be performed by the state either directly or under the meta-regulatory model, with direct involvement of the biotechnology industry itself, subject to government oversight.

2.4 The License Model

The License Model, developed by Gunningham, Kagan and Thornton (2003), seeks first to explain why large business enterprises behave the way they do towards the environment, before considering the normative and policy implications of such an analysis. It views business enterprises as simultaneously motivated and constrained by a multi-faceted 'license to operate', that includes not only the terms of their regulatory permits and legal obligations, but also an often-demanding 'social license', and a constraining 'economic license', which represent the demands of social and economic actors respectively.

These regulatory, economic and social license requirements are monitored and enforced by the stakeholders who generate them, and who commonly seek leverage by exploiting a variety of license terms.[14] For example, environmental groups not only enforce the terms of the social license directly (for example, through shaming and adverse publicity) but also seek to influence the terms of the economic license (for example, generating consumer boycotts of environmentally damaging products) and of the regulatory license (for example, through citizen suits or political pressure for regulatory initiatives). Thus the *interaction* of the different types of license often exceeds the effect of each license acting alone. The terms of some legal license provisions extend the reach and impact of the social license by directly empowering social activists, or by giving them access to information, or a role in the permit-granting process that they can use to pressure target enterprises. Conversely, a company that fails to respond appropriately to social license obligations risks a tightening of its regulatory license, as frustrated community activists turn for help to politicians and regulators.

However, the terms of each strand of the 'license to operate' are often far from certain or immutable, and are therefore subject to interpretation. What might be considered a 'reasonable' rate of return depends in part upon general market conditions and the particular investor. Different community groups can make conflicting demands. And different regulators can have different interpretations of the same set of environmental rules. Moreover, proactive corporate officials can sometimes reshape some license terms – by providing information for, and negotiating with regulators or environmental activists, by engaging in community outreach and education, by scanning for technologies and procedures that simultaneously cut costs and improve the firm's environmental performance. In effect, the influence of the regulatory, economic and social licenses on environmental performance depends on an 'intervening variable' – *managerial attitudes*, or the combination of attitudes and executive action called *environmental management style*. Management style is the perceptual filter through which management interprets its license to operate. Figure 1.1 below illustrates the License Model of corporate environmental behavior.

In the context of biotechnology, this model may provide insights that sharpen understanding of the interaction between different types of licenses and in particular between the regulatory and social licenses. For example, the particular strength of NGOs is in subjecting expert claims to public scrutiny, in challenging the risk-based and scientific-based nature of decision-making, and in injecting the values of sustainability, the precautionary principle and transparency, into decision-making (see, generally, Newell and Glover 2003). Their instruments for doing so: consumer boycotts, shareholder activism, alliances with supermarkets and other strategies described in terms of civil regulation above, will each be strengthened to the extent that the state intervenes directly with strategies of empowerment. For example, it might require greater transparency in the decision-making process, the provision of information (for example, labeling, which facilitates consumer sovereignty), opportunities for private litigation (class actions, rules of standing, strict liability versus negligence etc.) and it can facilitate various forms of public participation (for example, forums and independent review panels of experts).

The regulatory license also plays another crucially important role in curbing the food safety and environmental risks posed by biotechnology: bringing companies up to a minimum legal standard and in doing so, trumping the economic license, which otherwise drives biotechnology

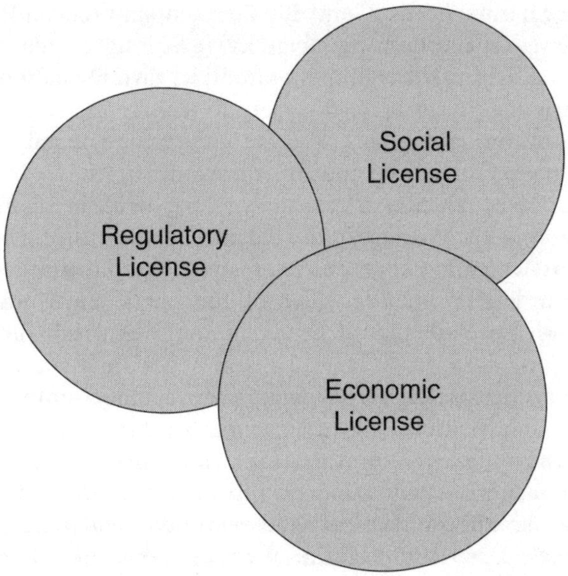

Figure 1.1 The License Model of corporate environmental behavior

companies to exploit the technology, free from constraints, in the interests of commercial advantage. Whether the regulatory license does indeed play such a role, will, as indicated earlier, depend substantially on the relative political strengths of the key stakeholders.

3 CONCLUSION

While each of the perspectives described above provides insights concerning how best to approach the task of regulating biotechnology, there are considerable disparities between them, and very different policy prescriptions flow from different perspectives. In terms of meta-regulation, the perceived solution is to establish regulatory structures that strengthen the capability of individual institutions or enterprises for internal reflection and self-control. For regulatory pluralism, it is a plethora of instruments that enable the state to steer not row, and to harness the capacities of second and third parties to fill more effectively the space vacated by the contracting regulatory state. From a civil regulation perspective, the state's principal role is to provide mechanisms that will empower the institutions of civil society to make corporations more accountable, and for the License Model, it is to devise mechanisms that reinforce the various strands of the social and economic licenses, and the interactions between them, with a particular focus on empowering social licensees.

A common theme in the regulatory pluralism, civil regulation and license frameworks is the power of civil society to change the behavior of large reputation-sensitive companies, who are vulnerable not only to shaming, but also to market forces and consumer pressure.

Each of the above frameworks has something valuable to offer and none of them are 'right' or 'wrong' in the abstract. Rather, they make differing contributions depending upon the nature and context of the policy issue to be addressed. The contrasting cases of 'first generation' environmental problems and the challenge of biotechnology regulation highlight this point.

Finally, we return to the role of direct state regulation under a next generation approach. We do so because it is important not to lose sight of the residual but nevertheless important role that command and control regulation can and should continue to play in environmental policy. It is only the state that can impose criminal sanctions and the full weight of the law, and only the state that, under statute, may have power of entry into private property to inspect, take samples, and gather evidence of illegality more generally. While there may be some circumstances where, as advocates of civil regulation, meta-regulation and regulatory pluralism would argue, far more can be achieved by various other forms of state and non-state action, this is certainly not the case across the board, and in the case of biotechnology, the role of the state is likely to remain central.

NOTES

1. There are of course, numerous other regulatory challenges posed by biotechnology, not least with regards to human genetics and patents, but these are beyond the scope of the present chapter.
2. The possibilities for divergent interpretation in such a highly politically charged and value-laden area are illustrated by the fact that while the US National Research Council has labeled GM foods as a safe food source, the British Medical Association, working off essentially the same facts, has called for an indefinite moratorium on GM foods (Prakash and Kollman 2003, p. 618).
3. Some, such as the regulatory flexibility initiatives established under the Clinton-Gore 'Reinventing Environmental Regulation' initiative, were directly inspired by one version of regulatory pluralism (and by Osborne and Gaebler's 1992 concept of 'steering not rowing' in particular). Seeking to embed environmental values and processes within the corporate culture in such a way that it becomes self-regulating, and relying upon oversight from local communities and perhaps third party auditors, to supplement or even replace direct regulation, is a quintessential pluralist strategy. Many informational regulation initiatives can also be understood in pluralist terms. Providing communities and financial markets with greater information about corporate environmental performance, effectively empowers both of these groups.

4. The StarLink episode in the US demonstrates how the principles of self-regulation can be flagrantly abused; see Bratspies (2003).
5. For a detailed analysis of the major components of effective regulation and a review of existing regulatory systems, see Jaffe (2004).
6. This, in Teubner's (1983) terminology, is a form of law using indirect means to achieve broad social goals, being a distinct shift from the previous approach (material law), which has broad goals but uses specific direct means to achieve these goals. Material law, according to Teubner, has largely failed because the complexity of modern society is incapable of being matched by a legal system of comparable complexity capable of harnessing direct, goal-seeking law, to accommodate social goals.
7. Defined as a form of law having broad goals and using specific direct means to achieve the goals.
8. For an excellent analysis of the alternative and much more reflexive regime that evolved in the aftermath of Three Mile Island see Rees (1994).
9. The term 'reflexive' derives from Teubner (1983).
10. This concept is further developed by Parker (1999a), Parker (1999b) and Braithwaite (1999).
11. Civil society is conventionally defined as involving 'citizens acting collectively in a public sphere to express their interests, passions, and ideas, exchange information, achieve mutual goals, make demands on the state, and hold public officials accountable. Civil society is an intermediary entity, standing between the private sphere and the state' (Diamond 1996).
12. The term used as a variant of civil society, *not* to imply the use of civil rather than criminal law.
13. For example, a highly successful Greenpeace campaign has been largely responsible for sensitizing European consumers (particularly in Germany and the UK) to the clear felling of old growth forests. This has had a profound impact upon North American companies exporting to those markets, who are increasingly being pressured by European buyers to provide evidence that the timber they supply has come from sustainably harvested sources.
14. Economic stakeholders include shareholders, customers, clients, suppliers and institutional investors and lenders. Legal stakeholders include regulatory, judicial and legislative officials at all levels of government: municipal, state and federal. Social stakeholders include local community groups, national environmental groups and, on occasion, the general public.

REFERENCES

Baldwin, R. (2004), 'The New Punitive Regulation', *Modern Law Review*, **67**(3), 351–83.

Bendell, J. (2000), 'Civil Regulation: A New Form of Democratic Governance for the Global Economy', in J. Bendell (ed.), *Terms of Endearment: Business, NGOs and Suitable Development*, London: Greenleaf, pp. 239–54.

Braithwaite, J. (1999), 'Restorative Justice: Assessing an Immodest Theory and a Pessimistic Theory', *Crime and Justice*, **25**, 448.

Braithwaite, John and Peter Drahos (2000), *Global Business Regulation*, Cambridge: Cambridge University Press.

Bratspies, R. (2003), 'Myths of Voluntary Compliance: Lessons from the StarLink Corn Fiasco', *Michigan State University College of Law Research Paper No. 01-07*.

Cullen, William Douglas (1990), *Piper Alpha Inquiry*, London: HMSO.

Diamond, Larry (1996), 'Towards Democratic Consolidation', in Larry Diamond and Marc F. Platter (eds), *The Global Resurgence of Democracy*, Baltimore, MD: Johns Hopkins University Press, p. 228.

Fiorino, D. (1999), 'Rethinking Environmental Regulation: Perspectives from Law and Governance', *Harvard Environmental Law Review*, **23**(2), 441–69.
Gamble, J.K. and C. Ku (2000), 'International Law: New Actors and New Technologies: Center Stage for NGOs?', *Law and Policy in International Business*, **31**, 221–62.
Gereffi, G., R. Garcia-Johnson and E. Sasser (2001), 'The NGO-Industrial Complex', *Foreign Policy*, July–August.
Grabosky, P. (1995), 'Using Non-government Resources to Foster Regulatory Compliance', *Law and Policy*, **17**(3), 256–81.
Gunningham, N. (1991), 'Private Ordering, Self-regulation and Futures Markets: A Comparative Study of Informal Social Control', *Law and Policy*, **13**(4), 297–326.
Gunningham, Neil A. and Peter Grabosky (1998), *Smart Regulation: Designing Environmental Policy*, Oxford: Oxford University Press.
Gunningham, Neil A., Robert A. Kagan and Dorothy Thornton (2003), *Shades of Green: Business, Regulation and Environment*, California: Stanford University Press.
Gunningham, N.A., M. Phillipson and P. Grabosky (1999), 'Harnessing Third Parties as Surrogate Regulators: Achieving Environmental Outcomes by Alternative Means', *Business Strategy and the Environment*, **8**(4), 211–29.
Jaffe, G. (2004), 'Regulating Transgenic Crops: A Comparative Analysis of Different Regulatory Processes', *Transgenic Research*, **13**, 5–19.
Koch, C. and K. Nielsen (1996), 'Working Environment Regulation: How Reflexive – How Political? A Scandinavian Case', Working Paper, Lyngby: Technical University of Denmark.
Murphy, D. and J. Bendall (1998), 'Editorial', *Greener Management International*, **24**, 8.
Newell, P. and D. Glover (2003), 'Business and Biotechnology: Regulation and the Politics of Influence', *IDS Working Paper 192, Biotechnology Policy Series 17*, Brighton: Institute of Development Studies.
Orts, W.E. (1995), 'Reflexive Environmental Law', *Northwestern University Law Review*, **89**(4), 1227–340.
Osborne, David E. and Ted Gaebler (1992), *Reinventing Government: How the Entrepreneurial Spirit is Transforming the Public Sector*, Boston, MA: Addison-Wesley.
Parker, Christine (1999a), *Just Lawyers*, Oxford: Oxford University Press.
Parker, C. (1999b), 'Evaluating Regulatory Compliance: Best Practice and Standards', *Trade Practices Law Journal*, **7**(2), 62–71.
Parker, Christine (2002), *The Open Corporation: Effective Self-regulation and Democracy*, Cambridge: Cambridge University Press.
Prakash, A. and K.L. Kollman (2003), 'Biopolitics in the EU and the US: A Race to the Bottom or Convergence to the Top?', *International Studies Quarterly*, **47**, 617–41.
Rees, Joseph V. (1988), *Reforming the Workplace: A Study of Self-regulation in Occupational Safety*, Philadelphia, PA: University of Pennsylvania Press.
Rees, Joseph V. (1994), *Hostages of Each Other: The Transformation of Nuclear Safety Since Three Mile Island*, Chicago: University of Chicago Press.
Teubner, G. (1983), 'Substantive and Reflexive Elements in Modern Law', *Law and Society Review*, **17**, 239.
Teubner, Gunther, Lindsay Farmer and Declan Murphy (eds) (1994), *Environmental Law and Ecological Responsibility: The Concept and Practice of Ecological Self-organization*, Chichester: John Wiley.

2. Rethinking regulatory governance for the age of biotechnology

Colin Scott

1 INTRODUCTION

The rapid development of biotechnology in the last 30 years, with applications to human health and reproduction, and to the agricultural, insurance and security sectors, has generated varied policy responses from governments in the OECD countries. Though often labelled 'biotechnology regulation' the vast bulk of the policy literature is concerned with the construction of only one element of a regulatory regime – the normative structure of principles, standards and rules. Biotechnology regulation, as a field of public policy, has not yet matured to the point where other elements of regulatory regimes – notably processes for monitoring and mechanisms of behavioural modification – are routinely considered or problematized.

This pattern of neglect of the machinery for implementation of regulatory policy is common to the emergence of earlier regulatory regimes – for example, occupational health and food safety (Paulus 1974). As with those earlier stories it creates risks that the norms, elaborated after much advice and discussion, may not be rendered effective in practice. On the other hand there is an opportunity to consider an array of mechanisms, going beyond traditional instruments of command and control, through which regulatory norms might be made effective. Such an analysis reveals not only variety in the mechanisms of regulatory implementation, but also highlights the possibility that norms agreed through state decision-making processes might be bypassed and even destabilized by alternative mechanisms for the generation and application of norms over biotechnology.

2 BIOTECHNOLOGY AND REGULATION

Biotechnology is not especially new. The fermentation of alcoholic drinks and the creation of cultures for making cheese and yoghurt are early applications that date back to antiquity. Similarly the selection of animals for breeding, and the development of new or better crops and flowering plants have occurred for many hundreds of years. Processes of genetic modification are extremely well established within human activities and accordingly might be thought to raise few entirely new ethical and political issues (Nuffield Council on Bioethics 1999).

Notwithstanding the long history of biotechnology there has been, in the last 20–30 years, a qualitative change in the policy debates, linked to concern and even panic about the potential application for new techniques for the engineering of plant and animal life, and in particular genetic modification (Ungar 2001). It is significant that the term biotechnology was coined as late as 1919.

This chapter provides an initial set of observations about the development of biotechnology regulation in response to these heightened policy concerns from the perspective of contemporary scholarship within the generic field of regulation. Within regulation scholarship there is a tendency to take policy objectives as given and to focus on the means by which they may be delivered – What types of rules? What types of regulatory institutions? How to monitor for compliance? What mechanisms of enforcement? (Baldwin and Cave 1999).

It seems to me that those embedded within the world of biotechnology policy have a nearly opposite set of preoccupations that focus on the development of the appropriate normative structures (an exception is provided by the regulatory analysis of Salter and Frewer 2003). Both approaches have weaknesses, of course. The latter risks neglecting how implementation of the policy might be made effective. The former tends to neglect the extent to which policy is made, implicitly or explicitly, through processes of implementation. Both approaches exhibit a bias towards privileging state norms and institutions – with an implicit assumption that governments can be and should be the main focus both for policy-making and implementation. An alternative perspective on regulation starts with the observations that in many and perhaps most regulatory domains the resources relevant to the capacity to regulate are widely dispersed among state and non-state actors, and that control within regulatory settings is not restricted to the development and enforcement of legal rules (Scott 2004). So, for example, whilst governments have a monopoly over the power to legislate, biotechnology firms possess much of the information that makes regulation possible, retailers exert considerable contractual power over food producers and, in some instances, non-governmental organizations (NGOs) possess some of the capacity to make regulation legitimate and to enforce either official or unofficial norms, for example, through direct action. The chapter does not claim to provide an authoritative analysis of the state of biotechnology regulation. Rather it is intended to use knowledge developed in the generic field of regulation to stimulate thinking about biotechnology.

The policy issues surrounding contemporary applications of biotechnology to human health and food production are concerned with five main sets of issues. The array of issues are concerned with safety of new technological applications (in particular in respect of food), the protection of the environment from irreversible change (in particular in respect of crops), the protection of consumers' economic interests (largely focussed on issues of disclosure and labelling), the protection of intellectual property rights and the complex ethical issues concerned with such issues as genetic testing, cloning and therapeutic applications of recombinant DNA.

3 BIOTECHNOLOGY AND THE FIELD OF REGULATION

The striking thing about this regulatory field is that so much of the public policy discussion focusses on epistemological issues concerned with what we know about these technologies and how we know it (Black 1998). In contrast with other fields of regulation a remarkable proportion of attention is devoted to the development of institutional structures for advising on and making policies and remarkably little to the machinery for implementing or enforcing the policies (Levidow et al. 1999). Thus, much attention is devoted to the important question as to what biases should operate in developing protective norms (and in particular whether to adopt the precautionary principle) but little attention is devoted to asking to what extent the protective norms are actually protective or to problematizing the machinery through which the policies are delivered.

We might go so far as to suggest that biotechnology policy has not yet matured into a field of regulation at all. In particular it does not seem to have encountered the severe doubts experienced in so many domains about the capacity of the state to exert public interested control over social and economic actors (Baldwin 1997). These doubts are expressed in many different ways. An early and well-known form of critique, sometimes labelled the Economic Theory of Regulation, focussed on an apparent tendency (particularly in the United States) for regulatory agencies to serve the interests not of the public, but rather of powerful private interests (Peltzman 1976). Sociologists of law have developed analyses of regulatory governance that focus on the communication problems between the various differentiated social sub-systems. This analysis questions the very possibility of deploying law for instrumental regulatory purposes (Teubner [1987] 1998). My own perspective on this particular issue is shaped by empirical work in which we have noted how widely dispersed are the resources of authority, information, wealth, organization and capacity to bestow legitimacy that make regulation possible among a variety of state and non-state actors (Scott 2001).

Anxieties about the capacity of what we might call 'classical regulation' to exert effective control has generated much discussion about the development of 'new instruments' for regulation. A recent analysis of new instruments in the environmental domain deployed a continuum comparing the extent of constraint imposed on firms, ranging between the highly constraining stick approach of classical regulation with the less constraining carrot approach of economic instruments such as subsidies, tradeable permits and taxes, with the least constraining sermon approaches that promote the dissemination of information as means to promote or shape choices in the market (Jordan et al. 2003).The promotion by the state of voluntary agreements features somewhere in this spectrum.

Casual usage of the term regulation often makes us think of regulatory agencies exercising control by reference to legal rules. In the biotechnology area much formal legal authority remains with government departments, rather than with agencies. Law is used:

1. to impose bans: for example, on human cloning and on commercial planting of GM crops;
2. to permit activities only within agreed limits: for example, controls over the use of the results of genetic testing by insurance companies;
3. to permit activities only with a licence;
4. to permit marketing of products only subject to certain conditions: for example, labelling of GM products.

We can see that these various uses of law take in the full range of the continuum of 'old' and 'new' regulatory instruments. This highlights a weakness in both the classical and the new instruments approaches to regulation, which is a relatively narrow focus on the conduct of government departments and agencies and on the deployment of law as setting normative standards or incentivizing compliance with such standards.

Given the anxieties raised by the development and application of biotechnology it is surprising that more has not been said about the capacity of classical regulation to exert control. It is not as if the biotechnology sector is without its regulatory fiascos. An American *cause célèbre* is the StarLink case. StarLink is a patented variety of genetically modified maize given regulatory approval in the United States for animal feed, but not for human consumption. A pressure group detected strains of the proteins unique to StarLink in taco shells marketed in the fast food outlets of

Taco Bell in September 2000. The manufacturer withdrew the products and subsequently other manufacturers detected similar proteins in their products and recalled them (Bratspies 2003).

The StarLink case is interesting because it demonstrates the failure to consider appropriate techniques for checking that restrictions imposed on the product were applied. The emergence of the product in human foodstuffs was not detected by a regulatory agency, but by a pressure group that proved itself to be the effective monitor for violations. The internal procedures of the various manufacturing firms involved had apparently not detected the problem. The consequences in terms of costs of recalls and for public confidence in regulation and in food markets were severe and international sales of American corn slumped when the relevant proteins were detected in shipments to Japan (Bratspies 2003).

The StarLink crisis was permitted to occur by the establishment of a regime of standard setting and authorization without paying attention to the methods of monitoring and enforcement (Bratspies 2003, p. 631). This is not to say that vigorous agency monitoring and enforcement was the only appropriate mechanism for addressing the issue, but rather that some attention has to be paid to the processes for implementing regulatory norms once they have been agreed. It is clear here that the state does not have a monopoly of power – there are major firms and interest groups with capacities for control. It is also apparent that the mechanisms that might have affected behaviour in this story are not restricted to legal oversight and might also include concerns with protection of reputation, transmission of emergent community norms and the interplay of forces in the market.

Regulation is more than just a set of norms. Cybernetic approaches to regulation suggest that a viable control system must have some goal, norms, standards or rules; some mechanism for monitoring or feeding back information about the performance of the regulatory system against the goal, norms, standards or rules; some means by which to modify behaviour that deviates from the required goal, norms, standards or rules (Hood et al. 2001).

Law's role in the monitoring and behaviour modification components of regulation is not unimportant. Here law is deployed to empower government agencies to engage in monitoring and collection of information and to empower departments, agencies, or courts to impose penalties or provide rewards for certain types of behaviour. But law also underpins self-regulatory regimes where firms come together to create binding rules and processes for themselves within an associational regime. The next section of the chapter offers an analysis of the full range of potential forms of control that might apply to any particular regulatory domain.

4 VARIETY IN CONTROL

Once we open up regulation into its components of goals, feedback and behavioural modification it is clear that in addition to the limited legal controls the regulation of the biotechnology sectors is, to varying degrees, subject to control through the pressures of competition and society. So, for example, rivalry between firms might be as important a factor in explaining the take-up and development of new technologies as a governmental regulatory regime. Such rivalry might even be harnessed for public purposes, for example, through a tendering process to determine which firm should win the contract to supply new seeds to a particular region's farmers.

The variation in the development of social norms relating to the development of GM crops appears to be a major explanatory factor in understanding the contrasting behaviour of firms in the United States and Europe. A striking example is provided by recent events in respect of GM maize in the UK. The UK government has indicated it is prepared to license GM maize for commercial crops, but Bayer, the main supplier of the technology, has decided to withdraw from the UK market. European governments at both EU and national level have taken a cautious approach to permitting the development of commercial GM crops, making reference to the application of the precautionary principle (Bernauer and Meins 2003). American frustration at the evidence that European consumers are even more cautious is spilling over into international trade litigation in which the right of European governments not only to ban GM crops, but also to require labelling of GM products (and thus inform doubtful consumers of which products to avoid) is being questioned (Prakash and Kollman 2003). Again the commercial actors appear at least as responsive to societal pressures as to governmental regulatory facilitation, as with Monsanto's decision to abandon its investment in herbicide-resistant GM wheat globally (Stokstad 2004).

Any adequate understanding of regulation should be able to accommodate the pressures based in rivalry and the application of social norms, capable as they are of bypassing the state altogether. To put the point another way, the biotechnology sector is subject to regulation whether or not the state does the regulating. Indeed, extensive standard setting by state agencies risks crowding out the capacities of businesses and civil society organizations to participate in or determine standard setting processes, and risks missing opportunities to promote 'ownership' of regulatory norms.

Attention to the diverse modalities of control is not restricted to the level or norms but also on the feedback and modification machinery. It is the disaggregated behaviour of buyers and sellers or of users and objects of competitive information in wider social settings that make competitive forces work. With social norms monitoring tends to be implicit, with social sanctions such as ostracization forming central mechanisms of behavioural modification (though community-based regulation can be formalized and linked to non-state hierarchy within self-regulatory regimes also). Key examples are provided within the application of biotechnology to human health. The machinery of public regulation is highly dependent on professional norms and monitoring within the medical and research communities (Kerr and Cunningham-Burley 2000). Inspection functions of the kind carried out by the UK Human Fertilisation and Embryology Authority (HFEA) are in essence a form of meta-regulation over the community-based control exerted through the professions.

Additionally, and acting on the insight of Lawrence Lessig that control can be built into the 'architecture' of products such as software (Lessig 1999), the behaviour of users of biotech products can be regulated by design features. The potential for using design to control behaviour has long been recognized in settings as diverse as prisons and Disney World (Shearing and Stenning 1985). A key commercial application of control-through-technology in the biotechnology sector is provided by the 'trait protection system' or 'terminator technology', which prevents the replanting of GM seeds from year to year, thus requiring farmers to make fresh purchases each year. The use of technology provides an alternative mechanism to legal enforcement to protect intellectual property rights in new products (Pendleton 2004, pp. 1–10).

It would, no doubt, be possible to build technological controls into GM crops for other than commercial purposes. Consequently, as with design features built into software to control the way it is used, design has at least the potential to form a basis for control. Indeed, technology-based controls could be used to promote regulatory compliance. It appears theoretically possible to build into GM products such as the famous StarLink corn, discussed above, features that would inhibit its marketing for purposes for which it is not licensed. In the case of StarLink if we assume that when animals are given corn (authorized) it is raw but that when it is prepared for humans (unauthorized) it is always cooked, it is possible to imagine accompanying the licence for growing and marketing the product with a requirement that technology is built into it such that it discloses itself as genetically modified when cooked. For example, it might turn blue after exposure to heat.

Thus, following Lessig's analysis of modalities of control (Lessig 1999) with adaptations (Murray and Scott 2002) it is possible to suggest that the hierarchical mechanisms associated with state law are but one of four possible modalities of control, the others being competition, community and design (see Table 2.1 below).

Table 2.1 Modalities of control in biotechnology regulation

Modality of Control	Regulatory Form	Biotechnology Examples
Hierarchy	Prior approval	Licensing (e.g. pharmaceuticals, GM crop planting) – controlling access to market
	Compulsory disclosure	Mandatory labelling (e.g. GM crops) – controlling information distribution
Competition	Market rivalry	Generic production of genetically engineered drugs – effects on price
Community	Public opinion – represented in direct action, spending decisions, etc.	Determining acceptability of commercial GM crop planting – affecting viability
Design	Built-in controls on use of products	Terminator technology in GM crops – controlling farmers' capacity to replant

These modalities are often not applied in pure form, but rather in combination. So, for example, hierarchical requirements relating to disclosure of information about GM foods are an important precondition to controls based both on competition and the application of community norms. The idea discussed above in respect of StarLink of designing technology-based controls into products so as to aid their regulation by hierarchical means is also a hybrid mode, combining hierarchy and design.

An objection to the analysis based in modalities of control might be to argue that relative to hierarchical control the other modalities are rather trivial. This may be true in some settings. But the field of biotechnology appears to be one in which the potential for control through community norms, through competitive pressures and through design does yield important examples. Hierarchical regulation may be rather more visible because it is formalized and institutionalized. This does not mean that it is always effective or that the explanation of how actors in a particular domain behave is explicable by reference to a regime of hierarchical regulation. The examples involving GM crops, discussed above, where governments have authorized commercial crops but firms have decided not to proceed, provide the clearest examples where, we might hypothesize, some other form of control is at play and is dominant in shaping the behavioural response. Indeed, in this case, some interplay between community norms and the market is undermining official government regulatory policy.

In some instances state regulation of one form or another underpins control based in competition or community or design. This is true in the case of licensing decisions underpinning competition through generic drugs and mandatory labelling underpinning consumer decisions on GM foods. But in other cases the application of modalities of control other than hierarchy appears to be spontaneous, in the sense that it was not initiated by the state for state purposes. In such instances it may or may not be right to refer to the mechanism of control as regulation, but it is surely impossible to wholly ignore such controls where they affect behaviour within regulatory regimes, in some instances enhancing the publicly set objectives and in other cases undermining them.

In addition to the descriptive objections to recognizing other modalities of control, there are also normative objections. It might be argued that if we subscribe to some version of the rule of law then it is wrong to permit other modalities of control to undermine public policy as it is set down in legal instruments. There are additionally moral objections to mechanisms of control that do not allow the subject control of choices as to their conduct (discussed in more detail in Roger Brownsword's contribution to this volume – Chapter 3). These objections apply most particularly to design-based control where the subject is physically inhibited from doing anything other than that permitted by the designer (Tien 2004). To a greater or lesser extent this objection, and the underlying argument that the subject is robbed of their responsibility for their actions by such mechanisms, applies also to control based in community or in competition, since participation in both markets and communities often requires us to accept their norms as a condition of participation, and non-participation may not be a viable option.

In the next section of this chapter we investigate the experience of hybrid forms of control in different regulatory settings and evidence as to the effects of such regimes. Central cases are likely to involve the combination of hierarchy with other modes of control. But equally interesting in the case of biotechnology regulation are instances where hierarchy is not present. We focus on experience of environmental regulation because of its proximity to some of the issues in biotechnology regulation.

5 EFFECTS AND EFFECTIVENESS

Stories of attempts by governments to exert control over social and economic activities are littered with accounts of counterproductive regulation (Grabosky 1995), fatal remedies (Sieber 1981) and catastrophes

(Moran 2001). Unintended effects are a pervasive consequence of human activity (Elster 1989). The focus in biotechnology regulation on the use of law to set norms, with the neglect both of other aspects of a viable regulatory regime, and of other modalities of regulation than law is liable to risk such unintended effects in these sectors too. The central questions addressed in this section are how to use the awareness of the potential of unintended effects and how to think about potential mixes of control types so as to maximize chances of producing intended effects.

Counterproductive effects in regulation are liable to result from a failure to anticipate how targets of regulation are likely to behave in the face of new regulatory requirements. The history of environmental regulation is full of such examples, for example, where rigorous new standards over vehicle emissions push up the cost of new cars to the extent that citizens retain older cars for longer, increasing rather than reducing total emissions. The development and marketing of GM foods provides an example of a product where the interplay of market behaviour and regulatory rules has led to effects within European markets that were not anticipated by governments.

Observations about variety in the availability of modalities of control alert us to the diffusion of both capacities and techniques by which regulatory effects are secured. Part of the problem of emphasizing state capacity and the use of legal rules in regulation is that it neglects or downplays not only the alternatives that might be available to do the job, but also the role that those alternatives will play in mediating whatever state regulation action is put forward in any case.

Putting the case positively, systems of control may emerge that are not wholly dependent on state action and legal rules, but which nevertheless effectively address the risks that policy-makers would want to address. In this section of the chapter I will make some observations about self-regulation in the biotechnology sectors and then consider regimes for environmental regulation that combine different modalities of control with apparent success. I have chosen regimes of environmental regulation to examine because they are longer established and better documented with empirical evidence than regimes for biotechnology regulation. But there are areas of overlap between biotechnology and environmental regulation and considerable potential for learning about the latter to inform regulatory policy in the former.

5.1 Self-regulation: The Case of Nuclear Power

Self-regulation is an important and long-established form of control, which draws on and develops norms within a community (for example, of professionals or businesses), often institutionalizing them and establishing mechanisms for monitoring and enforcement (Black 1996). A very well documented case involves self-regulation in the US nuclear power industry. Following the accident at Three Mile Island in 1979 there was global panic about the risks associated with the development of nuclear power as a key source for meeting future electricity needs. In many countries the cost and riskiness of nuclear energy, together with consideration of security issues, has led to the sector being established and retained within the public sector. In the United States, where the Three Mile Island accident occurred, there has been a long tradition of private ownership of electricity generation capacity including nuclear power. Though there have been public ownership experiments, notably in the Tennessee Valley Authority, private ownership has generally been the rule.

The characteristics of riskiness and moral panic attaching to the nuclear industry also affect some parts of the biotechnology sector. The industry response to Three Mile Island may be instructive. The circumstances of the industry response are beautifully captured in the title of Joseph Rees's book *Hostages of Each Other* (Rees 1994). The various firms within the US nuclear industry came to understand that the failure of any one of their number to act effectively on issues of nuclear safety would threaten the viability of the industry as a whole. Accordingly they developed the Institute of Nuclear Power Operations (INPO) as a system of self-regulatory standard setting and monitoring, which, in Rees's assessment, proved remarkably cohesive and effective. It was not an entirely freestanding self-regulatory system, since it depended for its enforceability on the actions of state regulatory agencies. Rees himself has referred to this aspect of the self-regulatory regime as 'the gorilla in the closet' (Rees 1997, p. 519).

So, the INPO story in the US is about the spontaneous response of an industry fearful of the consequences of inaction to the extent that the community of firms initiated tight controls that each was prepared to tolerate as the price for ensuring the future viability of the industry. The community-based controls were given added weight from the possibility of invoking the hierarchical capacity of state regulatory agencies. It is possible to argue that in some areas of the biotechnology industry the major market actors are similarly 'hostages of each other', in the sense that regulatory failures are liable to affect not just the firm responsible or whose products are directly affected, but the industry as a whole. The case is well illustrated by the StarLink affair, discussed above, where infractions of authorization (without any evidence of harm to human health) affected the whole international market for the supply of GM maize. The production of GM crops critically requires confidence in the standards of the market as a whole in a way that is not true of other potentially risky products such as cars. Put another way a major safety failure involving a Ford vehicle may have positive effects for the market position of General Motors and is unlikely to reduce the size of the car market overall. Major safety failures in the GM crops domain could kill off the entire market. Whilst the producers of GM crops will want to compete with each other, their viability is dependent on no safety scandal affecting any part of the industry. As with nuclear safety, cooperation and self-policing might be an attractive way to reduce the risks to the industry, though this will likely involve elements of governmental enforcement too. What about the possibility of a system of self-regulatory control that does not invoke the state's monopoly over legitimate force?

5.2 Combining Controls Based in Community and Competition: The Case of Forestry Standards

Environmental campaigners have long had concerns about the conduct of the logging industry in North America and elsewhere, and in particular with the problem of management of forestry and the environmental damage associated with excessive and inappropriate logging. The problems are global in scope, and it has proved very difficult to secure agreement on standards at the inter-governmental level.

An alternative response to national regulation or inter-governmental standard setting has come from the Forest Stewardship Council (FSC) established as a non-profit organization in 1993 (Meidinger 2003). The regime of control administered by the FSC extends to forestry in 60 countries and is a product of internationalization of environmental campaigning and linkages to the global marketplace. The FSC was established by NGOs, led by the WWF, as a mechanism for certifying good practice, but is recognized as legitimate by market actors (Cashore 2002, p. 507). The FSC regime combines interaction of the forestry industry and environmental campaigning movement in a hybrid form of control that draws on community- and competition-based forms of control – referred to by one commentator as non-state market-driven (NSMD) governance (Cashore 2002). The FSC rapidly spawned *industry-led* competitors in North America, which sought to develop recognition for alternative self-regulatory standards (Cashore 2002, pp. 508–9).

The normative structure of the regime derives from what are essentially community-based processes of interaction, institutionalized through the FSC, with market-based mechanisms. Indeed forest certification programmes 'recognize officially those companies and landowners who *voluntarily* operate "well-managed" or "sustainable" forestlands according to predefined criteria' (Cashore 2002, p. 505). The market element of the regime derives from the capacity of environmental groups to 'sell' the virtues of FSC certification to retailers as a positive symbol of retailer commitment to the environment and a mode of boosting sales. Retailers who accept the case are then able to use their economic power to contract only with suppliers who are FSC certified or at least to prefer such suppliers (Cashore 2002, pp. 509–10). At the supply end of the chain, firms that are seeking to maintain or extend market share and secure FSC certification are liable to apply the standards across all their activities, not just to the extent required to meet purchaser requirements in some countries. This rippling effect of the regime is an important feature and a product of globalization of markets.

5.3 Combining Hierarchy and Architecture with Community: The Case of Intentional Oil Pollution

An important question for the four-way analysis of controls and hybrid forms is to ask what does the design-based control add to the picture and to the potential for regulation. What is it that can be done through technological control that cannot be achieved through market, hierarchical or community-based controls? An important case that provides a partial answer to this question is provided by the history of attempts to prevent intentional pollution of the seas with oil. This is a domain in which there are many non-state actors with governance capacity, in addition to national governmental and inter-governmental organizations (Furger 1997).

The problem is that the operational practices of oil tankers involve the leaving of relatively small oil deposits after cargoes have been delivered. This 'clingage', representing 0.3–0.5 per cent of the original load, was routinely discharged to the sea before collecting the next cargo, either with sea water taken into the tanks to act as ballast for return voyages or as part of the process of cleaning tanks with high pressure sea water (Mitchell 1994, pp. 72–3). Intentional discharges accounted for a larger proportion of oil pollution to the sea than notorious accidental spills. The story of attempts to control intentional oil pollution at sea is one that demonstrates the weaknesses in depending on treaties made under public international law as mechanisms for promoting regulatory objectives. The key international treaty addressing this problem, the 1954 International Convention for the Prevention of Pollution of the Sea by Oil (OILPOL), was discredited by the 1970s because it proved very difficult to enforce discharge rules aimed at reducing pollution in the face of incentives by tanker operators to maintain their discharge practices. The weakness in the regime was identified as the human element – the requirement of a deliberate but virtually undetectable act of compliance (Mitchell 1994, p. 258).

Alive to the issues concerning intentional pollution the oil industry had developed two technologies geared towards its reduction. The first of these is to construct ships with segregated ballast tanks (SBTs) such that a separate ballast tank is created for sea water, which never carries oil, and does not require monitoring. Residues from loads ('slops') have to be retained separately on board and are subject to monitoring. Tank cleaning is addressed by a separate technology referred to as crude oil washing (COW), which uses oil to clean tanks, increasing the proportion of the load delivered (Mitchell 1994, pp. 78–9).The regulatory solution was for the International Convention for the Prevention of Pollution from Ships of 1973 and its 1978 protocol (MARPOL 73/78) to require the use of both technologies for new tankers with a reduction in the intentional pollution by four-fifths (Mitchell 1994, pp. 79, 258). For tankers already in service the convention required adoption of one or other of the technologies (Mitchell 1994, p. 258).

The MARPOL regime changed the point for regulatory enforcement from the moment at which a discharge is made to the point at which tankers were bought and put into service. As Mitchell points out, at this point compliance is dependent not only on the conduct of the tanker owner, but also the builder, a classification society (which certifies the quality of completed ships) and an insurance company (Mitchell 1994). In effect these other, non-state actors, become part of the regime for promoting compliance with the design standards. This non-state role is supported by the power of governments to detain tankers found to be in breach of pollution control standards. Thereafter the technology itself inhibits a substantial proportion of the targeted conduct. In any case, once the costly equipment is installed it is cost-effective to use it properly. Proof of the effectiveness of basing the regime on compliance with equipment standards rather than behavioural controls lies in data showing a very high degree of compliance with the SBT and COW requirements. By the mid-1980s nearly all tankers had both forms of technology on board. The critical difference between the two aspects of the regime was the way in which equipment standards were embedded within a configuration of actors capable of detecting violations and steering violators towards compliance and that compliance with the equipment standards then made non-polluting behaviour rational (Mitchell 1994, p. 292). Put another way the use of such equipment standards, through transparency and the involvement of many state and non-state actors, creates a system of 'redundant prevention' (Braithwaite and Drahos 2000, p. 288). Rules prohibiting discharge were targeting behaviour that tanker operators could engage in privately and with few opportunities for detection.

6 CONCLUSIONS

The field of biotechnology regulation does not yet seem to be troubled by the kind of anxieties affecting instrumental regulation in other policy domains. Yet there is already evidence in respect of GM crops in Europe that state regulation designed to promote confidence and thus create a viable market for the new technology has been undermined by a combination of direct action and the preferences of a doubtful public. Effective regulation of biotechnology applications in human health has not yet been so politicized.

It is difficult to avoid the thought that some policy fiasco or catastrophe is waiting to happen, and on a larger scale than the StarLink case. The domain has in common with the Enron case the necessity of a high degree of trust in core professionals and their capacity to develop and apply norms that support the public regulatory regime. This chapter is suggestive of a more holistic approach to biotechnology regulation in which the diverse capacities for control may be deployed to the benefit of the regime as a whole, rather than remain as neglected sources of vulnerability.

Thus where NGOs have an interest in setting standards, monitoring behaviour and realigning deviant behaviour they may become a central focus of a regulatory regime (as with the Forest Stewardship Council). Such NGO regimes of private regulation may be bolstered by the commitment of market players to use their contracting practices to recognize and give legitimacy to such regimes. This is particularly so where retailers and other big purchasers have higher brand recognition and greater public trust than suppliers. Market actors may also respond by establishing regimes in competition with NGOs, using purchaser choice as a mechanism for determining which regimes survive and develop by reference to which regime purchasers think will enhance their market positions.

Professional and other trade associations have also been demonstrably important in some environmental protection regimes, and clearly play key roles in some aspects of biotechnology, notably in medical applications. In some instances the concern for reputation of the industry as a whole may incentivize key actors to establish and develop regimes of standards, monitoring and enforcement that can effectively protect all players from the adverse consequences of a major regulatory failing or scandal.

None of this is to suggest that the state is not important. In some instances the mandating of technological controls is likely to be a governmental (or inter-governmental) function, as with controls of intentional oil pollution, which facilitates the operation of other modalities of control. In other instances, standard setting may occur through community mechanisms, but the capacity of the state to apply sanctions proves necessary to encourage compliance with those non-state standards, as with the US nuclear power story. Competition-based controls are liable to require some underpinning of the state, not just in competition policy in some instances, but also in the enforcement of contracts.

An intriguing question is whether we must envisage the state as always present in such regulatory settings, as a kind of meta-regulator over structures of control that have their own distinct orientation and mechanisms, but which the state can seek to fine tune to deliver public objectives. Alternatively is it likely that the incentives to operate controls over biotechnology regulation based in some combination of community, competition and design will leave the state with no necessary role, at least for some of the industries involved? This is one of the key issues for observers of regulation as the field of biotechnology evolves.

REFERENCES

Baldwin, R. (1997), 'Regulation: After "Command and Control"', in Keith Hawkins (ed.), *The Human Face of Law*, Oxford: Oxford University Press.

Baldwin, Robert and Martin Cave (1999), *Understanding Regulation*, Oxford: Oxford University Press.

Bernauer, T. and E. Meins (2003), 'Technological Revolution Meets Policy and the Market: Explaining Cross-national Differences in Agricultural Biotechnology Regulation', *European Journal of Political Research*, **42**, 643–83.

Black, J. (1996), 'Constitutionalising Self-regulation', *Modern Law Review*, **59**, 24–56.

Black, J. (1998), 'Regulation as Facilitation: Negotiating the Genetic Revolution', *Modern Law Review*, **61**(5), 621–60.

Braithwaite, John and Peter Drahos (2000), *Global Business Regulation*, Cambridge: Cambridge University Press.

Bratspies, R.M. (2003), 'Myths of Voluntary Compliance: Lessons from the StarLink Corn Fiasco', *William and Mary Environmental Policy Law Review*, 593–649.

Cashore, B. (2002), 'Legitimacy and the Privatization of Environmental Governance: How Non-state Market-driven (NSMD) Governance Systems Gain Rule-making Authority', *Governance*, **15**, 503–29.

Elster, Jon (1989), *Nuts and Bolts for the Social Sciences*, Cambridge: Cambridge University Press.

Furger, F. (1997), 'Accountability and Systems of Self-governance: the Case of the Maritime Industry', *Law and Policy*, **19**, 445–76.

Grabosky, P. (1995), 'Counterproductive Regulation', *International Journal of the Sociology of Law*, **23**, 347–69.

Hood, Christopher, Henry Rothstein and Robert Baldwin (2001), *The Government of Risk*, Oxford: Oxford University Press.

Jordan, Andrew, Rudiger K.W. Wurzel and Anthony R. Zito (eds) (2003), *'New' Instruments of Environmental Governance*, London: Frank Cass.

Kerr, A. and S. Cunningham-Burley (2000), 'On Ambivalence and Risk: Reflexive Modernity and the New Human Genetics', *Sociology*, **34**, 283–304.

Lessig, Lawrence (1999), *Code: and Other Laws of Cyberspace*, New York: Basic Books.

Levidow, L., S. Carr and D. Wield (1999), 'Regulating Biotechnological Risk: Straining Britain's Consultative Style', *Journal of Risk Research*, **2**, 307–24.

Meidinger, Errol (2003), 'Forest Certification as a Global Civil Society Regulatory Institution', in Errol Meidinger, Chris Elliott and Gerhard Oesten (eds),

Social and Political Dimensions of Forest Certification, Regmagen-Oberwinter: www. forstbuch.de.
Mitchell, Ronald B. (1994), *Intentional Oil Pollution at Sea*, Cambridge, MA: MIT Press.
Moran, M. (2001), 'Not Steering but Drowning: Policy Catastrophes and the Regulatory State', *Political Quarterly*, **72**, 414–27.
Murray, A. and C. Scott (2002), 'Controlling the New Media: Hybrid Responses to New Forms of Power', *Modern Law Review*, **65**, 491–516.
Nuffield Council on Bioethics (1999), *Genetically Modified Crops: the Ethical and Social Issues*, London: Nuffield Foundation.
Paulus, Ingeborg (1974), *The Search for Pure Food*, London: Martin Robertson.
Peltzman, Sam (1976), 'Toward a More General Theory of Regulation', *Journal of Law and Economics*, **19**, 211–40.
Pendleton, C.N. (2004), 'The Peculiar Case of "Terminator" Technology: Agricultural Biotechnology and Intellectual Property Protection at the Crossroads of the Third Green Revolution', *Biotechnology Law Report*, **23**, 1–29.
Prakash, A. and K.L. Kollman (2003), 'Biopolitics in the EU and the US: A Race to the Bottom of Convergence to the Top?', *International Studies Quarterly*, **47**, 617–41.
Rees, Joseph (1994), *Hostages of Each Other: The Transformation of Nuclear Safety Since Three Mile Island*, Chicago: University of Chicago Press.
Rees, J. (1997), 'The Development of Communitarian Regulation in the Chemical Industry', *Law and Policy*, **19**, 477–528.
Salter, B. and L. Frewer (2003), 'The Changing Governance of Biotechnology: The Politics of Public Trust', *Applied Biotechnology, Food Science and Policy*, **1**, 199–211.
Scott, C. (2001), 'Analysing Regulatory Space: Fragmented Resources and Institutional Design', *Public Law*, Summer, 329–53.
Scott, Colin (2004), 'Regulation in the Age of Governance: The Rise of the Post-regulatory State', in Jacint Jordana and David Levi-Faur (eds), *The Politics of Regulation*, Cheltenham, UK and Northampton, MA, USA: Edward Elgar, pp. 145–74.
Shearing, Clifford D. and Philip C. Stenning (1985), 'From the Panopticon to Disney World: The Development of the Discipline', in Anthony N. Doob and Edward L. Greenspan (eds), *Perspectives in Criminal Law*, Toronto: Canada Law Book Inc.
Sieber, Sam D. (1981), *Fatal Remedies: The Ironies of Social Intervention*, New York: Plenum Books.
Stokstad, E. (2004), 'Monsanto Pulls the Plug on Genetically Modified Wheat', *Science*, **304**, 1088–9.
Teubner, Gunther (1987), 'Juridification: Concepts, Aspects, Limits, Solutions', in Robert Baldwin, Colin Scott and Christopher Hood (eds) (1998), *Socio-legal Reader on Regulation*, Oxford: Oxford University Press.
Tien, L. (2004), 'Architectural Regulation and the Evolution of Social Norms', *International Journal of Communications Law and Policy*, **9**; http://www.digital-law.net/IJCLP/.
Ungar, S. (2001), 'Moral Panics Versus the Risk Society: The Implications of the Changing Sites of Social Anxiety', *British Journal of Sociology*, **52**, 271–91.

PART II

Regulating human genetics

3. Red lights and rogues: regulating human genetics

Roger Brownsword*

1 INTRODUCTION

This chapter presents some reflections on three related matters concerning the regulation of human genetics. The focal points for these remarks are, first, the nature of regulation itself; second, the respective influence of purchasers and providers in making a market for the products and services of the human genetics industry; and, third, the significance of viewing human genetics less as a regulatory target and more as a regulatory tool. The prompt for these reflections is what I take to be the standard test-case for the 'regulability' (Lessig 1999) of human genetics, namely the rogue geneticist intent on, say, engaging in human reproductive cloning and willing to carry out the research and development in whichever jurisdiction permits or, in practice, ignores such activities.[1]

The test-case of the rogue geneticist invites the following thoughts: that the essential regulatory challenge is to put in place effective global *prohibitions* against the abuse of human genetics; that rogue *providers* of human genetic products and services should be our principal regulatory targets; and that, if we cannot get a regulatory grip on such *target* rogue geneticists, our worst nightmare (say, the birth of a cloned child) will become a reality. The purpose of this chapter is not to suggest that these thoughts are wholly mistaken so much as to offer some correction for the particular loading that they reflect – that is to say, the loading for regulatory prohibition, for providers as the source of the regulatory problem, and for the fear that human genetics will prove to be a target that regulators simply fail to reach.

The chapter is in three sections. In the next section, section 2, I present a four-dimensional analysis of the nature of regulation, seeking a double correction, in part against thinking purely in terms of regulatory prohibitions (regulatory *red lights*) and, in part, against thinking narrowly in terms of traditional legal modes of regulation (red light *rules*). In section 3, the correction is again a double one: first, to draw attention to the role of *respectable* (not rogue) *providers*; and, secondly, to bring the activities of *purchasers* into view, the point being that markets for human genetic products and services will be made as much by purchasers as by providers. Where purchasers and respectable providers are so complicit, the burden of responsibility for creating regulatory problems should not be borne solely by rogue geneticists.[2] In section 4, following up a matter introduced earlier in section 2, I outline what is perhaps the most important correction. This is that, while unregulated (or unregulable) human genetics running out of control is one kind of nightmare, so too is the prospect of a regulatory regime that relies on the knowledge and understanding brought forward by human genetics to develop new 'techno-regulatory' instruments (Brownsword 2004a). In other words, we need to worry, in more or less equal measure, about the implications of human genetics escaping control as a regulatory target as about human genetics being co-opted as a regulatory tool.

2 THE NATURE OF REGULATION: A FOUR-DIMENSIONAL VIEW

What do we mean by 'regulation'? As Julia Black, among others, has pointed out, the concept of regulation is suffering from middle-age spread (2001, p. 103). For my purposes, let us assume that regulation covers whatever strategies are employed by an authorized legal agency to control or 'channel' human conduct in a particular field. In other words, we are equating regulation with one of Karl Llewellyn's law-jobs (Llewellyn 1940); we are also allowing that regulation can be a feature of simpler non-state communities; but, crucially, we are anticipating that there is more than one way of regulating. Just as alternative dispute resolution (ADR) has come to represent a range of alternative modes of dispute resolution (that is, alternative to third-party court-centred adjudication), we might speak of alternative modes of channelling (here, the alternatives being options other than rule-like legislation).

Consider, first, a simple regulatory example, that of channelling the actions of motorists at a crossroads, in order to avoid injury to person or damage to property. When traffic is light, it might suffice to channel through a rule of the road enjoining motorists to negotiate the crossroads with due care and attention. As traffic increases, motorists might find that, because they are unsure about the responses of other motorists, they need a rule that is more calculable. Hence, the due care standard is replaced with a rule that requires motorists to give way, say, to traffic approaching from the right. But this rule, too, proves unsatisfactory as motorists continue to lack confidence about one another's driving responses. At this stage, traffic lights are introduced and motorists are taught the significance of the red, amber and green lights. All is not perfect – some motorists, for example, 'jump' the lights and a few ignore them altogether – but, for the most part, the combination of rule plus lights channels the conduct of motorists as desired.

So far, so familiar. However, if the installation of traffic lights had not been an obvious response, consideration might have been given to a number of alternative regulatory strategies. For example, regulators might have considered running a public information and education programme relying on a better understanding of the crossroad code reinforced by the peer pressure and social disapproval of fellow road users. Or, they might have considered introducing a local tax for use of the routes leading to the crossroads, or a more general financial measure, in the expectation that some motorists would be priced out and the traffic reduced to a level at which the problem disappears. Or, they might have thought about a designed solution, such as changing the road layout from a crossroads to a roundabout or eliminating the intersection by building a flyover or underpass. Or, in a higher-tech future, they might have required manufacturers to design cars so that collisions are automatically avoided. None of these alternatives seems like a distinctively legal response. However, each represents a regulatory option; each seeks, in its own way, either to channel conduct so that the problem is reduced or to remove the problem by means of a technical fix.

At this point, we should pause. If channelling can be carried out in more than one mode, some rule-based, others technology-based, just how many dimensions are there to regulation? If rules and red lights are just one option, where do they fit in the larger regulatory picture? Let me suggest that we conceive of regulation as having four key dimensions, namely: phasing, pitch, mode and range.

2.1 Regulatory Phasing

Regulation might be 'first phase', 'second phase', or even 'third phase', and so on, in the following sense. Where regulation is first phase, its purpose is to control, confine and channel ex ante the particular aspect of practice (genetic or otherwise) that is its target. Where such first phase regulation is successful (genetic) practice operates (by and large) in accordance with the rules laid down by the regulatory order. Where, by contrast, regulation is second phase, no attempt is made to control, confine or channel some given aspect of (genetic) practice; regulators have abandoned such ex ante first phase intervention. Instead, second phase regulation operates ex post, endeavouring to compensate for, or adjust in response to, the consequences of a (genetic) practice that cannot be controlled by first phase regulation. If second phase regulation is not feasible, regulators might even delay intervention until a third phase, seeking to channel conduct in a context where the (third phase) consequences of the (second phase) consequences (themselves being the consequences of first phase deregulation) invite a regulatory response.

In this light, the well-known disagreement between Francis Fukuyama (who is ostensibly pro-regulation in relation to biotechnology) and Gregory Stock (who is anti-regulation) can be read as concerning the futility (or not) of first phase interventions and the necessity to grapple with second phase consequences (Fukuyama 2002; Stock 2002). For Fukuyama, it is not too late to attempt first phase regulation; for Stock, the writing is on the wall and we would do better to address the implications of the (inevitable) biotechnological revolution. So, for example, if technology for the genetic enhancement of our offspring were to be developed to the point where it was safe and reliable, but prohibitively expensive for most parents, whereas Fukuyama might favour first phase gatekeeper regulation designed to control access to the technology, Stock's advice might be to apply our regulatory resources to second phase correction of those adverse social effects that seem likely to flow from technology being available only to a privileged elite.

In what follows, it is probably simpler to assume that we are dealing with first phase regulation (although the analysis is equally applicable to second or third phase regulation).

2.2 Regulatory Pitch

The way in which regulation seeks to engage with its targets is of enormous significance. Essentially, channelling might adopt one, or more, of three pitches, namely: moral, practical and behavioural.

Where the pitch is moral, regulation seeks to engage with the target's moral reason. The regulatory purpose and position is presented as legitimate; compliance (where channelling takes this form) is put forward as a matter of moral obligation.

Where the pitch is practical, regulation seeks to engage with the target's practical reason (but not specifically the moral aspect of practical reason). What regulators rely on is the more diffuse claim that there is 'good reason' for compliance; and channelling will have the desired effect where targets share the sense of good reason (even if there is no one reason that targets uniformly recognize as good).[3] Accordingly, no particular reason within practical reason is privileged although, in practice, the practical pitch will often appeal to the self-interest, especially the economic interests, of targets.

Where the pitch is behavioural, no attempt is made to engage with either moral reason in particular or practical reason in general. All that matters is that the channelling device employed engages with the targets in such a way as to achieve the desired behaviour.

2.3 Regulatory Modes

The idea of regulatory modes takes us back to the kind of options that we were considering in relation to the control of traffic at the intersection.

In the context of information technology rather than biotechnology, Lawrence Lessig has developed a regulatory matrix that brings together traditional legal rules with social sanctions, markets and 'code' (Lessig 1999). The starkest contrast here is between law-like East Coast regulation and West Coast code. For our purposes, 'code' can be read broadly as covering any kind of technical fix (such as a chip that controls access to an internet site or a filter that blocks spam-mail) as well as modification to the physical environment (Kumar Katyal 2002).

If we were to recognize that the regulation of human genetics might employ more than one channelling mode, what would this imply? It would mean that over and above traditional law-like regulation, there are other options. If we are sceptical that law-like regulation is unlikely to control rogue geneticists, there might be yet other strategies. Those who argue in favour of the peer pressure of the international scientific community, codes of practice, self-regulation and so-called 'soft law' are already thinking such thoughts. It is not obvious how markets in the strict sense might constrain the rogue human cloners; but, if the market mode is broadened to cover a financial fix (and, after all, most market economies are macro-managed by channelling the behaviour of businesses and consumers through an array of financial fixes), the withholding of funding for embryonic stem cell research or therapeutic cloning can be seen as a relevant regulatory intervention. As for code, there is no obvious technical fix but the potential of this regulatory mode is a matter to be picked up later in the discussion.

When we combine what we have just said about regulatory modes with what we said earlier about regulatory pitch, we find that the legal and social modes tend to adopt a moral pitch as well as a practical pitch; the market or financial mode is straightforwardly practical in its pitch; but code or technical modes are largely behavioural in their pitch. This suggests that, when there is a turn to techno-regulation, there is a change in regulatory thinking. Instead of giving targets reasons for compliance, channelling pure and simple takes over. And, to this extent, it is misleading to treat the various regulatory modes as operating on the same plane; the topography of this dimension (of mode) only becomes apparent when it is viewed from the dimension of regulatory pitch.

2.4 Regulatory Range[4]

In principle, whichever regulatory mode (or modes) are deployed, the desired channelling might be negative, neutral or positive. Regulators might wish to channel conduct so that x is not done (negative channelling), or so that x is done (positive channelling), or so that agents have a choice between doing x or not doing x as they prefer (neutral channelling). Where traditional law-like regulation is employed, in its crudest form negative channelling presupposes a legislated rule that prohibits x; positive channelling a rule that requires x; and neutral channelling a rule that permits x. For example, in relation to the regulation of human genetics, legislation might prohibit reproductive cloning, permit therapeutic cloning and require (hypothetically) all citizens to contribute samples for a national genetic database. Regulators, thus, may show human geneticists a red light or a green light, but the complexity of the intermediate positions is not captured by an amber light – at any rate, if it is captured, it is only in the general sense of cautioning that there is a lot more to the regulatory range than simple red light prohibitions and green light permissions.

There is a great deal to be said about the regulatory range (even when, in the dimension of regulatory mode, we are thinking only about traditional law-like modes of regulation). However, it will suffice for present purposes to identify three variables within the range.

First, where regulation prohibits some genetic practice, it might sanction non-compliance in more than one way. Prohibition might signify the creation of a criminal offence with penal sanctions – as with the statutory prohibition on human reproductive cloning in the United Kingdom – but this is not necessarily the case. For example, if legislation is introduced to regulate genetic discrimination, and if the model of existing anti-discrimination schemes is copied, then the prohibition will be in the form of a statutory tort. This will mean that the principal sanction for non-compliance will be a private law claim for redress, possibly with some pressure for conformity applied by regulators having supervisory responsibilities. Or, again, where prohibitions form part of a licensing scheme such as that centred on the Human Fertilisation and Embryology Authority (HFEA), the sanctions for non-compliance may be internal to that licensing scheme (for example, the revocation of a licence).

Second, where the regulatory position permits some genetic practice, this might be accompanied by various reservations and qualifications (permission with negative reservation) or by various measures aimed at encouragement and incentivization (permission plus facilitation). When the HFEA is mandated to license research on human embryos if, and only if, such research is judged to be necessary, this signals that the permission is subject to negative reservation. However, negative reservation might also be manifested one step removed from the public law permission in private law. For instance, by refusing to treat 'designer baby' agreements as legally enforceable contracts or by rendering the tort regime more claimant-friendly in response to pharmacogenomic products, the background permission would be hedged with negative reservation. Importantly, however, the regulators might give the permission some encouragement by taking steps to facilitate the practice in question. Away from genetics, this regulatory approach has been very much in evidence in the steps taken to clear the way for the development of e-commerce (Brownsword and Howells 1999). In relation to genetics, the patent system is one of the keys to facilitation. For, whatever the ultimate raison d'être of patents, the working assumption is that the prospect of patent protection for inventive work encourages investment in research and development in genetics (Warren-Jones 2004). In this light, the debate about the patentability of inventive work around the human genome is not about prohibition versus permission but about permission subject to negative reservation versus permission plus facilitation. To the extent that the outcome is in favour of patentability, we have permission plus facilitation; to the extent that it is against patentability, we have permission subject to negative reservation (Nuffield Council on Bioethics 2002a).

Third, to understand the character of the regulatory position, we need to attend to the regulatory *mix* and the regulatory *tilt*. In the absence of a blanket prohibition under the criminal law or an unvarnished permission, the regulatory position is liable to mix elements of public and private law. To say that, in the UK, embryonic stem cell research is permitted is to say something rather significant – because, in many European countries, we have zones of regulated prohibition rather than regulated permission (Halliday 2004). Nevertheless, so far as the UK is concerned, there is a good deal more to know: we need to know on what terms such research is permitted, whether the products of the research (for example, techniques that are developed for deriving and purifying embryonic stem cells) are patentable, what sanctions apply if the informed consent of embryo donors is not obtained, whether the donors have any redress if the research is commercialized in a way that they did not expect, and so on. In other words, the full *mix* of regulation needs to be kept in mind. The *tilt* of the regulation also needs to be brought into account. To the extent that some aspect of genetic practice is not prohibited, it is permitted; to the extent that it is permitted, it is not prohibited. However, we need to know whether the regulatory tilt is towards prohibition or permission. If the former, ambiguities will be resolved in favour of prohibition and, similarly, where regulation is silent on a point, prohibition will be assumed as the default position.

Conversely, if the tilt is towards permission, it will be assumed that permission is the default position and silence, ambiguity or equivocation will be interpreted accordingly.

More could be said about the regulatory range but this will suffice. The essential point to take forward is that there is a great deal more to the regulation of human genetics than pressing the red light of law-like prohibition. Regulation should be conceived of in, at least, a four-dimensional way (Murray and Scott 2002). Accordingly, whether or not there is some permutation in the regulatory repertoire that will succeed in controlling rogue geneticists, we should not allow the test-case to distort our understanding of regulation. The test-case scarcely gets to even a one-dimensional view of regulation. By way of a corrective, it needs to be understood, first, that negative channelling is not always the desired effect; and, secondly, where it is desired, regulators have more than one mode in which to pursue their channelling objective, more than one way in which to pitch it, and more than one phase for the timing of their intervention.

3 ARE ROGUE PROVIDERS THE ONLY VILLAINS OF THE PIECE?

The test-case invites the assumption that rogue providers are uniquely responsible for the regulators' problems. If only there were no rogue providers, all would be well. I want to correct this idea in two respects. One correction is to see that *respectable* providers, too, might be a serious problem; and the other is to note that *purchasers*, as individuals or as members of a class, generate some of the regulators' difficulties.

With regard to the former, we can learn from Heidi Li Feldman's concept of 'markufacturing' or 'market-driven manufacturing' (MDM) (Feldman 2003). With regard to the latter, there is no shortage of illustrative cases – Diane Blood, the Hashmis and the Whitakers, all of whom have been involved in high-profile and controversial dealings with the UK Human Fertilisation and Embryology Authority, spring to mind; and we can learn, too, from the anticipated behaviour of insured persons who, knowing that they have a low-risk genetic make-up, wish to exit from insurance pools that are regulated by red light prohibitions against the use of genetic information.

3.1 The Problem of 'Markufacturing'[5]

Professor Feldman's starting point is that pharmaceutical companies are now under tremendous pressure to develop the next generation of major drug products. Patents do not last for ever; a good portion of patent protection is already mortgaged to finance investment in research and development; the majority of drugs entered for clinical trial fail to get beyond this stage; and, without new products, patents cannot be replenished. Patents and newly developed products thus form a 'virtuous circle'. In this setting, Professor Feldman predicts (or, should I say, already observes), an increase in the practice of product 'markufacturing'.

To explain, it is trite to say that a market will be moribund unless there is both supply-side and (correlative) demand-side activity. However, a market can be *initiated* on either the supply side or the demand side. In other words, a supplier may speculatively offer goods or services for which it turns out there is a demand; or a demand for goods or services may be articulated, to which a supplier then responds. What is distinctive about markufacturing is not simply that the market is supply-side initiated but that the supplier also, in effect, constructs the demand side of the market. In principle, we can imagine a spectrum of supply-side activity in relation to the construction of the demand side. At one extreme, the supplier simply puts the product into circulation and hopes that there will be a spontaneous interaction between product and demand-side interest of the kind that is required for the creation of a market. Some way in from this extreme, the supplier leaves less to chance, informing potential purchasers about the existence and the positive features of the particular product (for example, the advantages of audio-visual digital products over their pre-digital predecessors) and ensuring that, if there is a potential demand, it connects with the product to make a market. As we move further along this spectrum, marketing and advertising is intensified until we reach the markufacturing extreme, at which point the supplier manufactures both the product to be supplied and the demand-side need. When a product is markufactured, the supplier creates a culture around the product that convinces target purchasers that this is a product they need and must have. Putting the matter in the bluntest of terms, the markufacturing strategy of a pharmaceutical supplier is, first, to market the disease and then to promote the disease-responsive drug.

If we place markufacturing in the context of pharmacogenomics, we can anticipate that the marketing of tailored drug products will add a highly personalized gloss to the culture of need created by suppliers. Markufacturers will generate two mutually supportive messages: targets on the demand side will be persuaded not only that they have certain needs but also that some particular product is right for them – a case, as it were, of 'you *need* this' and '*you* need this'.

In its report on behavioural genetics, the Nuffield Council on Bioethics picks up on the demand-side aspects of markufacturing (without using this terminology) by highlighting the phenomenon of so-called 'diagnostic spread' (2002b). By this, the Council means that there is a tendency to medicalize conditions that, hitherto, have been regarded as lying within the normal range of human behaviour – for example, as the Council remarks, 'the producers of new "anti-shyness" drugs, such as Paxil and Luvox, have been accused of applying to normal behaviour, interventions developed for pathological traits' (Nuffield Council on Bioethics 2002b, para. 13.19). What is more, individuals who think that they might have General Anxiety Disorder or Social Anxiety Disorder can take an online self-test, which sets a suitably low threshold for professional advice to be recommended. The ingredients of this cautionary tale are replicated by the move to promote impotence drugs, such as Viagra, as a response to female sexual anxieties (or, as the drug companies wish to characterize it, the newly recognized medical condition of 'female sexual dysfunction', Boseley 2003a). In both cases, we find that once-normal anxieties are re-classified as pathological and, with the backing of commercial and social pressure, medical intervention is advised.

Of course, there is nothing new about markufacturing as such. The cultivation of consumer need for products, many inessential, some very damaging, is part of the history of consumer society. Nevertheless, for two reasons at least, markufacturing by the pharmaceutical companies in the information society is a cause for concern. One reason is that the revolution in information technology facilitates the collection of data about purchasing patterns, which then aids and abets extremely targeted marketing. The other reason is that markufacturing by pharmaceutical companies might not only trade on a degree of trust but also take advantage of human vulnerability leading to the consumption of products that might have seriously adverse effects.

Professor Feldman's thesis is that the law should respond by reflecting the special responsibility of markufacturers for the markets they create and for the damage that they might cause. To this end, some upgrading of the tort regime seems to be called for; or possibly regulators need to get a grip on markufacturing before it can stimulate demand and dependency. However, this is not a debate that we need to enter. For our purposes, it is enough to note that markufacturers, who are in no sense akin to the rogue geneticist of the test-case, are likely to contribute to the concerns of regulators.

3.2 Purchaser Power

If respectable providers will contribute to the regulators' difficulties, the same applies to respectable purchasers. Here, we are not presupposing purchasers whose needs have been created by providers. We are thinking about individuals with 'normal' instincts. In the first two illustrative examples, the instinct is for the flourishing of one's family; in the third example, it is simple economic rationality.

3.2.1 Diane Blood

Diane Blood's campaign to have access to her deceased husband's sperm so that she could start the family that she and her husband had planned is a story that has been told many times (Adams and Brownsword 2003, p. 108). It does not concern human genetics as such; it is a particular episode in the reproductive revolution ushered in by the development of in vitro fertilisation (IVF) techniques. However, Diane Blood's story does tell us something about the persistence of determined individuals; the limited resources of regulatory bodies when issues are litigated; and the way in which public sympathy for reasonable campaigns can be mobilized (Human Fertilisation and Embryology Authority 2003).[6]

The lessons of *Blood* are reflected in the Human Fertilisation and Embryology Authority's report on sex selection. Public opinion runs strongly against access to sex selection techniques for social purposes;[7] but the Authority recognizes that determined couples will seek out the required services in jurisdictions that lie beyond its regulatory reach (Human Fertilisation and Embryology Authority 2003, para. 130). For sure, we can agree that global markets for reproductive services coupled with determined purchasers will add up to sales. Moreover, if the media is able to mobilize support for a woman having access to her deceased husband's sperm so that she can have his children would it be so difficult to mobilize support for a couple who already have a houseful of girls and who would now like a boy? The Authority's survey of public opinion suggests that there is little sympathy for such a couple but it would be extremely surprising if such a cause (for the sake of a balanced family) could not be spun as eminently reasonable in a carefully cultivated context.

3.2.2 The Hashmis and the Whitakers

The plight of the Hashmi family has attracted widespread interest. Briefly, one of the Hashmis' children, Zain, was born with the blood disorder, beta thalassaemia major. To rectify this disorder, Zain needs a stem cell transplantation, using the cord blood or bone marrow of a donor who is free of the thalassaemia gene as well as being a tissue-match with Zain. None of Zain's three elder siblings being a tissue-match, the Hashmis resolved to have another child (an intended saviour sibling). The first child conceived by Mrs Hashmi was aborted because prenatal tests showed that it had the thalassaemia gene; and the second, although born healthy, was not a tissue-match for Zain. In desperation, the Hashmis sought to have access to the latest preimplantation genetic diagnosis (PGD) and tissue-typing procedures in order to select a suitable embryo for implantation.

Before such a procedure could be lawfully undertaken in the UK, it needed to be specifically authorized by the Human Fertilisation and Embryology Authority (the Authority). For, although there was nothing new about PGD being carried out as part of IVF treatment licensed by the Authority,[8] tissue-typing had not previously been carried out as part of such treatment.[9] In December 2001, the Authority announced that, in principle, it would be prepared to license tissue-typing but only where PGD was already being licensed for a serious genetic disorder. Such licences (for PGD plus tissue-typing) would be granted only exceptionally – namely, where the particular genetic disorder was severe or life-threatening and where the embryo itself was at risk of this disorder; provided also that all other possible avenues for treating the affected child (such as Zain) had been explored; that the intended recipient was not a parent; that the intention was to use only the cord blood; that the parents would be counselled; that the family would be encouraged to participate in follow-up studies; and that no embryo would be genetically modified to provide a tissue match. In a blaze of publicity, in February 2002, the Authority gave the go-ahead for the Hashmis.

The efforts of the Hashmis (who had undergone two rounds of treatment without success) were interrupted when Comment on Reproductive Ethics (CORE), a group for whom absolute respect for the embryo is axiomatic, sought a judicial review of the Authority's December 2001 announcement and, by extension, the licence granted the following February. CORE argued that the Authority had no power to issue a licence permitting the use of human leukocyte antigen (HLA) typing to select between healthy embryos.

At the time of writing, the Court of Appeal has ruled that the Authority acted within its powers, but the case is likely to go on to further appeal. Whatever its outcome, it illustrates the determination of individuals to access the latest reproductive technology.[10] And that is the point that matters for present purposes.

Alongside the Hashmis' successful application to the Authority, there is the equally relevant case of the Whitaker family. The Whitakers, too, need a saviour sibling, for their son Charlie who suffers from the rare blood disorder, Diamond Blackfan anaemia. The Whitakers, like the Hashmis, sought authorization for a tissue-typing test. However, the Authority refused because the test would not be carried out in conjunction with PGD – this particular blood disorder not being one for which there is a genetic marker (and thus not a disorder that would be identified by PGD). Accordingly, the Whitakers were forced to go (perfectly lawfully) to the

United States for the tissue-typing but without a covering licence from the Authority. The perseverance of the Whitakers paid off when they returned to the UK to celebrate a successful donor birth (Boseley 2003b).

There is a complex ongoing debate about whether the line drawn by the Authority between these two cases is defensible in principle (Brownsword 2004c, pp. 316–17). However, the story of the Whitakers not only underlines that of the Hashmis', in the sense of exemplifying the determination of purchasers, it adds to it in an important way. For, if determined purchasers (such as the Whitakers) can get what they want by entering markets outside the jurisdiction, local regulatory controls are seriously weakened. Even if the line drawn by the Authority is defensible in principle, it will be difficult to hold in practice. And, as we shall see shortly, this phenomenon of purchasers going round local restrictions to access desired products of the genetic revolution is not limited to reproductive or medical services.

3.2.3 The case of insurance and genetic information

In *Genes and Insurance*, Marcus Radetzki et al. argue against red light regulatory regimes ('total regulation') that prohibit insurance companies from either requiring genetic tests to be taken or, where test information is already available, requiring disclosure of the results (2003). The authors' principal contention is that such regimes are unlikely to have the intended effect of shielding those whose genetic make-up would otherwise disadvantage them in the insurance market; and, indeed, the consequences of adopting such regulatory approaches might be so counterproductive that the market itself collapses when it can no longer bear the weight of its high-risk burden.

Let us assume that the intended purpose of such regimes is to prevent insurance companies from discriminating against applicants whose genetic profile shows up as a bad risk and, concomitantly, to protect such applicants against being priced out of, or excluded altogether from, vital insurance or insurance-financed services. In this way, the intention underlying such regulation is that the private insurance market should mimic solidarity-based social insurance. Yet, the authors observe, the regulators seek to do this at just the time that social insurance is being dismantled. Not only this, this particular regulatory approach is adopted in the face of burgeoning information about the human genome and in a context of globalization. Putting these pieces together, the authors argue that red light regulation simply will not work.

To understand why there is a problem with such regulatory regimes, we need to recall a truism about disclosure in contracting situations. In many contractual negotiations (including negotiations for insurance), there is an asymmetry of information. The asymmetry might favour the purchaser or the provider. If the former, then the rational-economic purchaser will disclose the information in question if the effect of disclosure will be to reduce the price. However, if the effect of disclosure will be to increase the price, the rational-economic purchaser will not disclose. Now, imagine the position of an applicant for insurance who has information (asymmetrically) about his own genetic make-up. If disclosure of that information will exert a downward pressure on the premium, the information will be disclosed; but if disclosure would exert an upward pressure, the applicant would not usually wish to disclose. The intended effect of total regulation is to permit the latter kind of applicant to withhold the information and, thus, to enjoy what is in effect a discounted premium (as well as avoiding whatever collateral disadvantages might flow from putting this information into circulation). However, the secondary effect of total regulation is that low-risk applicants, who are willing to disclose, are either not permitted to do so or are not able to profit from doing so. In consequence, lower-risk applicants (unwillingly) subsidize those who are higher-risk and, with this, there is a danger that the pool will be destabilized through the process of 'adverse selection'.

In the first instance, the stability of an insurance pool (comprising both high-risk and low-risk insured parties) subject to a regime of total regulation depends on the attitudes of the three key stakeholders. Those who are high-risk parties will normally wish to sustain the pool on total regulation terms (because it is to their economic advantage to do so); those who are low-risk will wish to revise the regulatory terms on which the pool operates so that premiums or contributions paid by individuals more faithfully reflect real risk profiles or, failing this, they will wish to exit the pool in order to join a pool with more favourable terms; and, depending upon a number of economic variables, the insurance companies might or might not prefer the pool to be run under total regulation terms or on terms that allow for premiums to be set on sophisticated differential terms. Given this, although there might be some occasional abuses of a total regulation scheme by those who know that they are high-risk, the major impulse for change (and the most likely source of destabilization) will come from low-risk stakeholders who would prefer to participate in a pool where premiums are adjusted in line with disclosed risk profiles.

Whilst the authors' thesis that total regulation invites adverse selection is certainly plausible, whether or not low-risk stakeholders actually leave the pool depends upon a number of variables about which we can be rather less certain. First, there is the question of how much genetic information we have and just how informative such information actually is in telling an insurer about the risk that an applicant represents if covered by a particular type of insurance. Much depends on whether we posit a cautious or a bold scenario with regard to the development of our genetic understanding – the more that we assume a bold scenario, the greater the significance of the suppressed information, and the greater the pressure for disclosure.

Second, for low-risk applicants to have access to insurance pools that are regulated in accordance with standard rules for disclosure, there must of course be pools of this kind and applicants must be able to access them. If we assume a bold scenario in relation to the development of genetic understanding, then we can reasonably expect that there will be a sufficiency of low-risk demand to move insurers (not necessarily the same insurers who operate in total regulation schemes) to supply in response to this demand. On these assumptions, we can assume that, in a global marketplace for insurance, local regulators will have great difficulty in preventing their citizens from having access to such schemes.

A third variable concerns the extent to which low-risk parties display inertia in sticking with their existing pools. Again, the assumption of a bold scenario invites the speculation that even sluggish consumers of insurance will be moved to transfer to another pool in order to cash in their low-risk advantages. Admittedly, there needs to be a degree of confidence about the prospects and security of the new pool; but, in 10 to 15 years' time, consumers should be perfectly comfortable with shopping around for insurance services that are distance-sold in electronic environments.

Fourth, there is the question of whether insurers who operate in total regulation schemes *and who have an interest in sustaining such schemes*, can either impose costs against exit that deter low-risk members from leaving or devise incentives for staying within the pool. For example, if there are significant premium discounts for insured parties who keep their policies, low-risk and high-risk, with a single total regulation insurer, the drift away from such schemes might be arrested. Finally, other things being equal, short-term economic considerations encourage low-risk members to move out of total regulation schemes. However, if the longer-term effect of this drift away from such schemes is to jeopardize the viability of insurance clubs that serve high-risk members, then those persons who are both low-risk (for one type of insurance) and high-risk (for another) might see that their short-term calculation is in tension with their longer-term reliance on high-risk pools.

Having said this, it seems reasonable to assume that, other things being equal, low-risk members will take their business away from insurance pools that are subject to total regulation. As a pool progressively becomes a club for high-risk members, premiums necessarily will rise – and, indeed, as insurers understand what is happening to the pool, they might well inflate premiums in anticipation of the higher risks that they now realize they are covering. With membership becoming more expensive, even high-risk insured parties might have to drop out; and, in due course, dwindling membership might threaten the viability of the scheme so that the few remaining members are treated as uninsurable.

Tourism, it seems, whether to procure medical services or to join more favourably regulated insurance pools, will be a constant headache for local regulators – and it is purchasers as much as providers who are responsible for giving the regulators a hard time.

3.2.4 Taking stock

What do these tales of 'heroes' and 'villains' tell us? Quite simply, they alert us to the fact that the consumers of genetic products are not merely passive. Purchasers are active; they will seek out markets, sometimes heroically so, that are regulated in ways that they find attractive; and the knock-on effects of their market-seeking activities might well be felt by their own local regulators – at which point the heroes become villains. The rogue geneticist might create problems for the regulators but we should not generalize to assume that all the regulators' problems will emanate from either rogues or respectable providers; purchasers, too, whether as ground-breaking heroes or collective villains, will play their part.

4 REGULATORY TARGET OR REGULATORY TOOL?

Thus far, we have tended to think about human genetics as a regulatory challenge rather than a regulatory opportunity, as a regulatory target rather than a regulatory tool. Our test-case confirms this one-directional thinking. In this fourth section of the chapter, I want not only to correct against this perspective but to consider the consequences if we invert it. Suppose that, instead of viewing the rogue geneticist as an outlaw, beyond regulation, regulators were to take on board the new genetics as a regulatory instrument. Cloning, no longer a regulatory problem, becomes a regulatory solution. How could we possibly arrive at such a perverse view? Let us go back to where we are starting, with the test-case of the rogue geneticist, with the concern that regulation will prove ineffective.

4.1 Regulatory Attitudes: From Defeatism to Perfectionism

According to Francis Fukuyama (2002), 'pessimism about the inevitability of technological advance is wrong, and it could become a self-fulfilling prophecy if believed by too many people. For it is simply not the case that the speed and scope of technological development cannot be controlled'. On the other hand, Fukuyama concedes that:

> [N]o regulatory regime is ever fully leak-proof, and if one selects a sufficiently long time frame, most technologies end up being developed eventually. But this misses the point of social regulation: no law is ever fully enforced. Every country makes murder a crime and attaches severe penalties to homicide, and yet murders nonetheless occur. The fact that they do has never been a reason for giving up on the law or on attempts to enforce it. (2002, p. 189)

Allowing that idealism should not drive us to pessimism, the question remains: in the real world, how should regulators respond to the explosion of the new genetics? One response is, in effect, to abandon regulation (resigning ourselves to a regulatory race to the bottom and the prospect of technological might dictating what is right). A second response is to try at least to hold the regulatory line, concentrating resources on the most serious violations.[11] A third response – the one that is especially pertinent for our purposes – is to turn to new technology to assist with regulatory projects. To respond in this way, regulators would not need to be perfectionists who aspire towards zero-tolerance and total compliance (although this might help). All that it would take would be the birth of a breed of regulators, willing to invest heavily in regulatory technologies, and to think creatively about how technology might improve regulatory effectiveness.

4.2 The Regulatory Learning Curve: Towards Techno-regulation

Regulators surely will learn from their experience of seeking to regulate the new technologies. They will understand more about the strengths and weaknesses of traditional forms of regulation as they will come to appreciate the potential of employing the new technology itself in a regulatory role – not merely to regulate the primary users of the technology (in the way, for example, that privacy-enhancing technology might be deployed in information technology) but far more broadly (in the way that genetic profiling, closed circuit television (CCTV), computer mapping of crime, monitoring and tagging, and so on are already used in the criminal justice system). The speculation is that, alongside traditional forms of regulation, we will find new forms of technologically assisted regulation being piloted and adopted.[12]

Initially, technology is, and will be, employed within the framework of traditional legal modes of regulation. The technology might be designed to discourage non-compliance or to improvc the chances of detection, or both; it might be pretty crude (for example, speed bumps or other traffic-calming measures within restricted areas) or it might be more sophisticated (for example, closed circuit television (CCTV), smart cards, tracking devices, forensic databases, and so on). No matter how sophisticated the technology, however, there is always the option of non-compliance and always some chance that one will not be detected.

Next, the technology is improved to the point where, although the regulatory mode remains legal, non-compliance will be detected and the application of the designated sanction is guaranteed. We can imagine an 'all-seeing' 24/7 surveillance technology that enables the regulators to monitor our every action. If we fail to comply, we will be seen – and, in conjunction with database and recognition technology, we will be identified. We can elect non-compliance but we do so in the certain knowledge that we will pay (Norris and Amstrong 1999, Chapter 10).

The decisive step comes, however, when regulators make no attempt to engage with practical reason, when legal, social and financial modes give way to 'techno-regulation'; when rules and regulations give way to technologically secured results; in the ideal-typical case, when compliance is guaranteed because non-compliance is not an option. How so? Possibly, this could be by a technical fix to the environment – for example, by taking cash as we know it out of the economy and replacing it with electronic money (which, admittedly, might simply replace one kind of crime with another) – or by a technical fix to human biology (employing programmes of genetic screening and selection) or by a combination of controlling interventions applied to human predisposition and our particular environment. With techno-regulation, there is no such thing as the perfect crime; criminality is no longer an option.[13]

4.3 The Nightmare that is Techno-regulation

In the lead up to techno-regulation, we can anticipate that there will be concerns about the regulators' invasions of privacy (Gallagher 2003; Gentleman 2003; Norris and Armstrong 1999; Whitfield 1997) – including, as Geoff Peck's recent successful action before the European Court of Human Rights (ECtHR) illustrates all too vividly,[14] their casual handling of personal data once collected and stored (O'Neill 2002, pp. 109–10). If the all-seeing all-knowing regulatory state is necessary for the protection of more compelling rights, some loss of privacy might be justified. However, serious questions about the limits of technological assistance will be raised, for the context required if a community of rights is to flourish can only take so much intrusion. Once we enter the realm of techno-regulation, however, the emphasis shifts not simply to the efficient elimination of crime but to the elimination of choice, to the treatment of subjects as though they lack the capacity to choose. If a genetically engineered crime-free zone could be created, would we want to enter? Would this not represent the most fundamental kind of affront to human dignity, throwing doubt on the status of humans as bearers of both rights and responsibilities?[15]

Yet, this is such a blunt warning that it merits double-checking. Let us suppose that some action, x, is categorically contrary to human rights and human dignity. There is simply no merit in x being permitted. Let us suppose that the only way of ensuring that x is not done is to techno-regulate it. On these premises, how can it be sensible to oppose a techno-regulatory strategy to eliminate x?

First, regulation that engages with practical reason respects human dignity by giving individuals the choice of compliance or non-compliance. Regulation is not neutral between, or indifferent towards, these options. Compliance is very definitely the preferred option. However, the final choice is left to individuals. Accordingly, it is implicit in this model that human dignity values not only the right choice being made (to comply) but the process of choosing itself. Generalizing this, human dignity is committed to a framework for action in which humans may choose to do the right thing as they may choose to do the wrong thing; to take away from humans their capacity to make wrong choices is an insult to their capacity for choice, the worst kind of affront to their dignity.

Second, it might be objected that, even if techno-regulation could be justifiably used in the hypothesized case, it should be opposed as setting in motion a culture of diminishing respect for the importance of choice. The argument would run that, once it is accepted that it is legitimate to techno-regulate x where x is plainly wrong, it will not be long before it is accepted that it is legitimate to techno-regulate x where x is almost certainly, or probably, wrong; and so on down a slippery slope. The further we slide, the greater the risk that we mistakenly restrict options for action, and the greater damage that we do to the ideal of human dignity. Moreover, if this is a plausible risk in a well-intentioned human-rights respecting liberal society, how much greater the risk of abuse in a society where the ruling class discovers the full potential of its techno-regulatory powers? In the face of such risks, the argument is that we would do better to steer well clear of techno-regulation – it is a temptation to be resisted.

Third, it might be conceded that techno-regulation would be justifiably used in the hypothesized case – or, at any rate, it would be justifiable as a measure of last resort. However, it would be argued that the premises underlying the hypothesized situation are implausible and atypical. The crucial premise is that x is categorically contrary to human rights. How many such xs are there that would be applicable in a broad spectrum of contexts? If the answer is 'relatively few', then techno-regulation would be of little practical significance. Similarly, if many such (prima facie) xs can be found but they are context-dependent, then the option for non-compliance might be too difficult to define and design out of existence – in which case, techno-regulation would be of little practical significance.

Of these three arguments, it seems to me that, although the third points to some difficulties in putting techno-regulation into practice, it is the first two that represent the real lines of resistance. Of these two, the second is weakened by its reliance on contingencies; essentially, it is an invitation to precaution that some might decline. The main line of defence, therefore, is the first argument. Pushing this defence a step further, we approach Jürgen Habermas's view that human dignity requires a context in which agents are what they are, and responsible for their actions, by virtue of genetic chance and individual choice. In other words, human dignity requires a particular balance of chance and choice that must be respected if instrumentally rational interventions (for more effective crime control, for more effective humans) are not to become self-destructive.[16]

If this line of thinking is sound, the test-case to trouble us is not so much that of the rogue geneticist as that of instrumentally rational techno-regulators who abandon a regulatory 'republic of reason' (Sunstein 2001, pp. 239 et seq.) in favour of preventive unconditioned channelling.

5 CONCLUSION

I started out by suggesting that we need to correct against the bias of the test-case of the rogue geneticist. This test-case invites a limited understanding of the nature and range of regulatory options as well as a one-sided perception of the source of regulatory problems. In a sense, these reflections neutralize one another: the regulatory repertoire is richer than the test-case assumes but the sources of regulatory problems are more diverse than it suggests. Hence, we have greater regulatory strength than the test-case gives credit for; but this is just as well, because there will be many sources setting challenges for the regulators.

With the unfolding of these reflections, however, correction has taken on a more anxious tone. For, in the regulatory repertoire, there is the option of channelling conduct by giving it a technical fix. The more that technologies associated with research in human genetics are brought forward, whether by rogue geneticists or respectable scientists, the greater the attraction of this option; the technical fix will tend to become the default strategy. Techno-regulation might be as far off as genetically engineered humans; but we should not allow the test-case deviance of the rogue geneticist to throw us off the trail of what might be the most potent threat to human dignity.

NOTES

* I am indebted to the Leverhulme Trust, without whose support it would have been a great deal more difficult to have completed work on this chapter.

1. The reference to 'rogue geneticists' should be taken broadly to cover 'mavericks', 'cults' and 'cowboys' and the like, characterized by the mainstream as practising bad science, and/or governed by bad ethics, and/or guided by bad motives. At the time that I sketched out this chapter, the announcement that a group in South Korea had cloned a human embryo from which stem cells had been extracted prompted another round of concerns of this kind (Rose 2004).
2. Interestingly, Rose (2004) concludes that '[t]he problem of medical tourism by would-be parents is trivial compared with the need to control the search by biomedical researchers for countries with soft standards'.
3. Compare Habermas (1999), p. 330: '[T]hough modern law requires from its addressees nothing more than norm-conformative behavior, it must nevertheless meet the expectation of legitimacy, so that it is at least open to people to follow norms, if they like, out of respect for the law.'
4. This discussion of the regulatory range draws on some ideas first published in Brownsword (2004b).
5. In this part of the chapter, I am drawing on Brownsword (2003a). Although Professor Feldman uses the term MDM in the final version of her paper, I prefer the evocative 'markufacturing' that she used in earlier versions.
6. For the litigation, see *R v Human Fertilisation and Embryology Authority*, ex parte *Blood* [1997] 2 All ER 687. For comment, see Morgan and Lee (1997), p. 840.
7. In her Chair's Foreword to the report (Human Fertilisation and Embryology Authority 2003), Suzi Leather says that the 'strength of opposition to sex selection for any but the most serious of medical reasons . . . is unmistakable'. For example, 82 per cent of respondents to the consultation questionnaire *disagreed* with the statement 'The use of sperm sorting should be permitted in sex selection for non-medical reasons'; 82.85 per cent *disagreed* with the statement 'The use of PGD [preimplantation genetic diagnosis] should be permitted for selecting the sex of offspring for non-medical reasons'; 82.68 per cent *disagreed* with the statement 'Sex selection (by either sperm sorting or PGD) should be permitted for non-medical reasons when a family has at least two children of one sex and none of the other sex' [that is, sex selection for family balancing reasons]; and 85.06 per cent *disagreed* with the statement 'Sex selection (by either sperm sorting or PGD) should be permitted for non-medical reasons other than family balancing'.
8. The framework legislation, the Human Fertilisation and Embryology Act 1990, does not deal explicitly with licensing PGD. However, the Authority takes the view that because (1) the Act explicitly permits research on embryos, (2) the possibility of developing methods for detecting gene or chromosome abnormalities in embryos prior to implantation was recognized in 1990, and (3) a clinical trial to develop PGD for a life-threatening sex-linked disorder had just been undertaken at the time, the Act implicitly supports the licensing of PGD for severe or life-threatening disorders (Human Fertilisation and Embryology Authority and ACGT 1999, para. 10).
9. Neither the Human Fertilisation and Embryology Act 1990, nor the Authority's own (then current) Code of Practice (5th edition, 2001), deals explicitly with the question of tissue-typing.
10. See *R (Quintavalle on behalf of Comment on Reproductive Ethics) v Human Fertilisation and Embryology Authority* [2002] EWHC 2785 (Admin); [2003] EWCA 667 Civ. For comment on the Court of Appeal's decision, see Brownsword (2004c, p. 304). (The appeal did go on to the House of Lords where it was rejected. See [2005] UKHL 28.)
11. Cf. Perri 6, (2002, p. 145), for a dispassionate analysis of what we can reasonably expect from our regulators.
12. Recent suggestions include the use of scannable microscopic computer chips embedded in high-value bank notes (to combat forgery); vehicle number-plate recognition

technology to enable cars to be matched against an insurance database (to combat driving without insurance); joined up information systems (following the murder of Victoria Climbié, to identify children at risk); and tagging paedophiles using global satellite positioning technology. On the first, see Osborn (2003); on the second, Coughlan (2003); on the third, Cross (2003); and, on the fourth, Doward (2003). The introduction of 'smart' ID cards is another important indicator of this tendency: see Travis and Wintour 2003).

13. Cf. Lessig (2001, p. 249):

    Technology, tied to law, now promises almost perfect control over [artistic] content and its distribution. And it is this perfect control that threatens to undermine the potential for innovation that the Internet promises.

    To resist this threat, we need specific changes to re-establish a balance between control and creativity. Our aim should be a system of sufficient control to give artists enough incentive to produce, while leaving free as much as we can for others to build upon and create.

14. *Peck v United Kingdom* (2003) 36 EHRR 719. The Court held that the actions of a local authority in releasing CCTV footage that captured a suicide attempt by Peck engaged his Article 8 right under the Convention. And, whilst the local authority may have been well-intentioned in its desire to publicize the success of its CCTV scheme, its failure either to mask Peck's identity or to obtain his consent meant that its interference with his Convention right was disproportionate and unjustified.
15. Here, I am presupposing a conception of human dignity that underpins a will theory of (human or agency) rights, Beyleveld and Brownsword (2001); Brownsword (2003b); Brownsword (2003c).
16. As Habermas (2001, pp. 92–3), puts it:

    The morality of egalitarian universalism stands in question as such. To be sure, this modern form of moral consciousness provides the only rationally acceptable basis for the normative regulation of action conflicts in pluralistic societies. But why shouldn't complex societies simply drop their normative foundations entirely, and switch over to systemic (!) (or, in the future, biogenetic) steering mechanisms? . . . Today, the relevant controversy is played out between a naturalistic futurism, committed to a technical self-optimization of human beings, and anthropological conceptions whose 'weak naturalism' has them accept the views of neo-Darwinism (and scientific views in general) without scientistically undermining or constructivistically outstripping the normative self-understanding of speakers and actors, for whom reasons still count.

REFERENCES

Adams, John N. and Roger Brownsword (2003), *Understanding Law*, London: Sweet and Maxwell.

Beyleveld, Deryck and Roger Brownsword (2001), *Human Dignity in Bioethics and Biolaw*, Oxford: Oxford University Press.

Black, J. (2001), 'De-centring Regulation: Understanding the Role of Regulation and Self-regulation in a "Post-regulatory" World', *Current Legal Problems*, **54**, 103–46.

Boseley, S. (2003a), 'Drug Firms "Invent" Sex Disorder', *The Guardian*, 3 January, p. 8.

Boseley, S. (2003b), 'As Age of the Saviour Sibling Dawns, Pressure Mounts Inexorably to Change Embryo Rules', *The Guardian*, 20 June, 3.

Brownsword, R. (2003a), 'Causes for Concern and Causes of Action: A Comment on "Pushing Drugs"', *Washburn Law Journal*, **42**(3), 601–14.

Brownsword, R. (2003b), 'Bioethics Today, Bioethics Tomorrow: Stem Cell Research and the "Dignitarian Alliance"', *Notre Dame Journal of Law, Ethics and Public Policy*, **17**, 15–51.
Brownsword, R. (2003c) 'An Interest in Human Dignity as the Basis of Genomic Torts', *Washburn Law Journal*, **42**(3), 413–78.
Brownsword, Roger (2004a), 'What the World Needs Now: Techno-regulation, Human Rights, and Human Dignity', in Roger Brownsword (ed.), *Human Rights, Global Governance and the Quest for Justice, Volume IV*, Oxford: Hart, pp. 185–202.
Brownsword, R. (2004b), 'Regulating Genetics: Dilemmas of Form and Substance, Doctrine and Design', *Medical Law Review*, **12**, 14–39.
Brownsword, R. (2004c), 'Reproductive Opportunities and Regulatory Challenges', *Modern Law Review*, **67**(2), 304–21.
Brownsword, R. and G. Howells (1999), 'When Surfers Start to Shop: Internet Commerce and Contract Law', *Legal Studies*, **19**, 287–315.
Coughlan, S. (2003), 'Driving Home the Hot-spot Danger Zone', *The Guardian* (Jobs and Money), 13 September, 6.
Cross, M. (2003), 'Eyes on the Child', *The Guardian* (Life), 18 September, 16.
Doward, J. (2003), '500 Paedophiles to be Tracked by Satellite Tags', *The Observer*, 21 September, 1.
Feldman, H.L. (2003), 'Pushing Drugs: Genomic and Genetics, the Pharmaceutical Industry, and the Law of Negligence', *Washburn Law Journal*, **42**(3), 575–99.
Fukuyama, Francis (2002), *Our Posthuman Future*, London: Profile Books.
Gallagher, C. (2003), 'Nothing to Hide, Nothing to Fear?', *Liberty*, 6.
Gentleman, A. (2003), 'ID Cards may Cut Queues but Learn Lessons of History, Warn Europeans', *The Guardian*, 15 November, 21.
Habermas, J. (1999), 'Introduction', *Ratio Juris*, **12**(4), 329–35.
Habermas, Jürgen (2003), *The Future of Human Nature*, Cambridge: Polity Press.
Halliday, S. (2004), 'A Comparative Approach to the Regulation of Embryonic Stem Cell Research', *Medical Law Review*, **12**, 40–69.
Human Fertilisation and Embryology Authority (2003), *Sex Selection: Options for Regulation*, London: HFEA.
Human Fertilisation and Embryology Authority and ACGT (1999), *Consultation Document on Preimplantation Genetic Diagnosis*, London: HFEA.
Kumar Katyal, N. (2002), 'Architecture as Crime Control', *Yale Law Journal*, **111**(5), 1039–140.
Lessig, Lawrence (1999), *Code and Other Laws of Cyberspace*, New York: Basic Books.
Lessig, Lawrence (2001), *The Future of Ideas*, New York: Random House.
Llewellyn, K.N. (1940), 'The Normative, the Legal, and the Law-jobs: The Problem of Juristic Method', *Yale Law Journal*, **49**, 1355–400.
Morgan, D. and R.G. Lee (1997), 'In the Name of the Father? *Ex parte Blood*: Dealing with Novelty and Anomaly', *Modern Law Review*, **60**(6), 840–56.
Murray, A. and C. Scott (2002), 'Controlling the New Media: Hybrid Responses to New Forms of Power', *Modern Law Review*, **65**(4), 491–516.
Norris, Clive and Gary Armstrong (1999), *The Maximum Surveillance Society: The Rise of CCTV*, Oxford: Berg.
Nuffield Council on Bioethics (2002a), *The Ethics of Patenting DNA*, London: Nuffield Council on Bioethics.
Nuffield Council on Bioethics (2002b), *Genetics and Human Behaviour*, London: Nuffield Council on Bioethics.

O'Neill, Onora (2002), *Autonomy and Trust in Bioethics*, Cambridge: Cambridge University Press.

Osborn, A. (2003), 'Computer Chip Plan to Fight Banknote Fraud', *The Guardian*, 9 June, 4.

Perri, 6 (2002), 'Global Digital Communications and the Prospects for Transnational Regulation', in David Held and Anthony McGrew (eds), *Governing Globalisation*, Cambridge: Polity Press.

R v Human Fertilisation and Embryology Authority, ex parte *Blood* [1997], 2 All ER 687.

Radetzki, Marcus, M. Radetzki and N. Juth (2003), *Genes and Insurance*, Cambridge: Cambridge University Press.

Rose, H. (2004), 'Beware the Cowboy Cloners', *The Guardian*, 16 February, 16.

Stock, Gregory (2002), *Redesigning Humans*, London: Profile Books.

Sunstein, Cass R. (2001), *Designing Democracy*, Oxford: Oxford University Press.

Travis, A. and P. Wintour (2003), 'ID Cards are on the Way', *The Guardian*, 12 November, 1.

Warren-Jones, A. (2004), 'Patenting DNA: A Lot of Controversy Over a Little Intangibility', *Medical Law Review*, **12**, 2–13.

Whitfield, Dick (1997), *Tackling the Tag: The Electronic Monitoring of Offenders*, Winchester: Waterside Press.

4. An abstract approach to the regulation of human genetics: law, morality and social policy

Justine Burley

1 INTRODUCTION

For all the talk about what is 'ethical and moral' in policy deliberations over applications of human genetics, many law-makers have proved themselves remarkably adept at eschewing the sort of abstract reasoning that would lend point to either term. Typically, they are deployed merely to signal that a given policy references a certain domain of human conduct or belief, not to add substance to a discussion or a clause in a document. The net result of this reluctance to engage with the issues presented by advances in genetics at a more critical level of our moral thinking, is that regulation is piecemeal, short-sighted, so effete in some cases that it does not serve the intended purpose, and in others so defective (morally or logistically) that it has actual or foreseeable adverse consequences for both individuals and society.

These are strong claims, to be sure. One of the aims of this chapter is to substantiate them. The case studies I shall concentrate on to that end, are genetic testing and insurance, and therapeutic cloning. Discussion of each case study will be carried out in dedicated sections, where I shall provide analysis of current debate, and of the benefits and pitfalls of actual legislation/regulation. I then go on to demonstrate further the ways in which existing measures are defective by recourse to a thought experiment involving a hypothetical world in which already-possible genetic technologies have advanced significantly, though still within the bounds of what is reasonably predictable. It is in this part of the discussion where I sketch a positive case for how we might approach, in the abstract, the construction of a regulatory framework by marshalling a conception of the two core values of modern liberal political philosophy. Thus, the second aim of the chapter is to demonstrate how abstract thinking can sensibly inform discrete policies, even if expediency requires that we depart from the abstract in practice. I judge that reasoning about the regulation of genetic technology from the angle just mentioned provides important insights to policy-makers.

Two important preliminary matters must now be attended to. I shall sketch the specifics of this fantasy scenario, and justify its use. I shall then spell out the two key values of mainstream political philosophy, upon which I later build and draw, with a view to assisting regulators in devising coherent principled regulation of applications of genetic technology.

2 LOOKING TO THE FUTURE

It is my view that no robust set of regulations for genetic technology can be framed by limiting the focus of inquiry to the here and now. Hence our inquiry will take into consideration a hypothetical 'What if?' scenario of a world in which genetic technology has advanced beyond the current state of the art. In this hypothetical world, genetic testing is accurately predictive for a vast number of monogenic and polygenic genetic disorders and predispositions. These tests are available in clinics, and many of them also by mail-order for home use. Genetic profiling of newborns is now current, affordable, and is instructive about their future state of health. Embryonic stem (ES) cell-based therapies enable the effective treatment of disease and tissue repair. The unproblematic creation of embryos using somatic cell nuclear transfer (SCNT) allows these to be carried out in patient-specific fashion. Pre-implantation genetic diagnosis (PGD) is an effective method of determining and eliminating which embryos possess single genes for deleterious, early and late onset diseases, and is employed to select embryos for implantation that have relatively superior polygenic combinations, in terms of predispositions to disease. Although I assume that all of the aforementioned are a reality, I shall also assume that *they are only permitted and practiced in a handful of nations.* (I ignore completely the discussions that took place in those countries in the lead up to the development and use of these technologies.) The decisions policy-makers face in those countries where the technologies are not permitted is whether to allow their introduction, and, if they decide in favor, what the shape of regulation should be. What considerations will be most pertinent to the decision-making process?

Why bother incorporating the fantasy scenario into discussion of the regulation of human genetics? First, it is incumbent on us to promote regulations that will not serve to endorse morally indefensible applications of a technology or deny morally permissible ones, in the future. What we lay down now by way of regulation, if reasoned poorly, may well run counter to the interests of humans in the medium term. We need, in short, to fix clearly on the moral direction of what matters. We might add that genetic science is not static. If the next 50 years bear witness to developments on a par with those of the last half-century, broad perspective is indicated given that the fantasy may, in whole or part, become a reality.[1] The state of genetic and related technologies in the fantasy world I envisage is, after all, one in which existing technology is more efficacious (genetic testing of various kinds) or has moved beyond proof of principle (ES cell research involving SCNT).

Second, advances in human genetics and other areas of bioscience already oblige us to confront problems that, at heart, are not morally meaningfully different than those suggested by the hypothetical scenario posited above. The thought experiment is intended to lend point and crispness to contemporary debate; debate that is currently muddied by sloppy thinking, skewed focus and the influence of competing interests. It enables us to distinguish the arguments that are the most crucial *at the level of application*. This is because it permits us to shelve debate about which improvements in the technologies will occur, and when they will occur, how safe and efficacious a technology will be, which moral slides down which slippery slopes (for example, the perfection of SCNT leading to successful human reproductive cloning) are threatened, etc. I do not for one moment wish to suggest that such issues are neither pressing nor important. They are both. However, the thought experiment helps us to see that the moral landscape changes somewhat when our thinking on the subject is liberated from thoughts about what is and is not likely to develop, efficacy, and so on. The implications of this for current regulations regarding both genetic testing in the insurance context, and therapeutic cloning, are arresting, as I shall show. It is highly doubtful that there will be an enforceable international consensus on whether and how genetic technologies should develop and/or be regulated. These decisions will ultimately reside, as they do now, with individual nations. Nations with currently unpermissive policies with regard to therapeutic cloning, and which, for example, prohibit the use of genetic testing for anything other than clinical or research purposes, may find in the not too distant future that their legislation falls nothing short of being indefensible on a number of important counts.

The thought experiment is revealing but it does not itself offer any principled moral argument. To develop my case more fully – that many existing policies regulating genetic testing and insurance, and therapeutic cloning (as well as ES cell research more generally) are effete, misguided and counterproductive – it is necessary to identify the core values at stake, and to apply these consistently in order to ground our discussion in principled foundations. In this way, the conclusions I draw about appropriate regulation can resist the charge that they have been plucked from thin air.

3 THE TENETS OF GOOD GOVERNMENT

What is the proper role of government in a liberal democracy? The most cogent answer to this question appears to have eluded many of those involved in drafting now-current/impending legislation/regulatory policy circumscribing applications of genetic technology. The answer I shall give to this question, by no means a novel one, strongly hints why recent law-making endeavours relating to genetics have resulted or are likely to result in problematic outcomes. It also points a way forward for enlightened regulators.

Two values form the core of contemporary liberal theories (Dworkin 2000, Chapters 2 and 3 and Rawls 1971). Both have helped to shape the legislative framework of actual liberal democracies. The first is that the state shows adequate regard for citizens when it creates the conditions for individual freedom, and respects those conditions by not seeking to promote *particular* ways of life such as adherence to a specific religious doctrine or sexual mores (whether or not a majority of citizens would consent to either), through policy. This regard for citizens is complemented by redistributive mechanisms designed to mitigate inequalities between individuals' natural capacities (state of health, genetic make-up, levels of talent). The liberal dual emphasis on freedom and equality informs any thorough assessment of existing legislation concerning human genetics, and also the abstract underpinnings of a sound regulatory framework. We are on firmer footing if we can find fault with policy initiatives relating to genetics, which run foul of these values.

Why are liberal theorists preoccupied with establishing the conditions for individual freedom? Why, for example, would they decry any attempt on the part of a state to impose, say, a certain set of religious beliefs on its citizens? The liberal view is premised on the fact that every single individual has his or her own unique conception of what it means for a life to go well – which projects, plans and attachments for each one of us, give life meaning. Even in relatively culturally homogeneous societies, pluralism of the kind just mentioned is prevalent. This begs the question of why the fact of pluralism is attached so readily to the normative idea that it is the role of government to be responsive to said differences between individuals. Part of the answer resides in the notion that the authors of all of these differing conceptions of the good – projects, associations, religious beliefs, and so on – are better placed than government to form, reflect on and revise them. According to the argument from expediency, it would be pointless were the state to insist that we all think and act in strictly circumscribed ways, when it could not possibly enforce these.

The other normative premise of the liberal approach to freedom is that whatever else its content, human well-being, at the baseline, consists in self-authorship, self-direction and self-governance. The capacity for autonomy distinguishes us from non-human primates, and from automatons. To deny human individuals the opportunity to be autonomous is to compromise their well-being. Thus, on the liberal account, government rightly furnishes the context for these baseline properties of well-being. It does so by guaranteeing certain rights and freedoms, for example, security of the person and property, religious worship, association, expression and equal political participation. Liberty on this account is not license. People are not free to do whatever they please, rather they are free co-extensively with others – if my religious or political view involves physically harming another human being, or censoring opinions in conflict with my own, then my liberty can be justifiably constrained. But, when my exercise of liberty does not directly impede the exercise of critical liberties by another, I am free to believe in and act on whatever I deem valuable.

It is worth emphasizing that the liberal understanding of the importance of freedom to human well-being requires that *strong reasons* be adduced before the state can legitimately constrain individual action. The mere fact that a majority of citizens might think themselves superior to a minority or think their views, religious or otherwise, superior to those held by a minority, is not adequate justification for any liberal state to favour that majority or to promote its views through policy. The value of freedom just described is the cornerstone of liberal democracy. It is the basis of a vast body of law, not to mention the basis of the economic, political and social systems of modern liberal societies.

The second value – equality – is premised on the fact that well-being consists of possession of an adequate and fair share of resources. If individuals are to have the opportunity to pursue their chosen ways of life they require more than freedom. People cannot eat freedom, nor can freedom enable a person who has a certain disability physically to perform certain jobs either at all or as easily as others can. Inadequate means of subsistence (nutrition and shelter), poor health and paucity of talent, in themselves, are held to be bads.[2] The normative premise of the liberal approach to equality is that the existence of relative differences between individuals' monetary resources and natural capacities is unfair.[3] Individuals' health, talents, wealth and income are often the product of brute luck[4] in the social and/or genetic lottery. Liberal theorists recommend redistributive mechanisms such as taxation on the grounds that in a socially interdependent world, it is reasonable to demand that the lucky demonstrate concern for the less fortunate. Individuals have long accepted this demand (though they do so unevenly and inconsistently). Actual liberal governments formalize it through progressive taxation and the various state social, health and welfare programmes that taxation funds.

The job of policy-makers and regulators is to respect the axioms of liberal government. They are, to repeat, to facilitate pluralism by establishing and maintaining the conditions for individual freedom, which entails a hands-off approach by government so long as no violation of a person's or group's critical liberty is at stake, and to mitigate human suffering and unfairness traceable to brute luck with policies targeted at achieving a fair distribution of a society's resources. The proverbial devil, of course, is in the detail, in how we work out policy specifics from a theoretical conception of these values. That being said, reference to even these general features of liberalism is instructive to discussion of existing legislation governing genetic technology, and of the design of future regulations. As I will demonstrate below, the liberal conception of both values is too often lost sight of in the policy-making arena as regards certain applications of genetic technology. It is not simply that the axiomatic features of good liberal government – respect for individual freedom, and redress of brute luck – have not been faithfully adhered to. One expects a departure from ideals in practice. What one does not and should not expect is that regulation flies in the face of these values. With this conception of the proper role of government, and with the story of the hypothetical world close to mind, I now turn to discussion of our case studies.

4 GENETIC TESTING AND INSURANCE

There is an impressive body of literature devoted to the description and analysis of a host of ethical, legal and social issues raised by genetic testing. It is not my intention to provide exhaustive treatment of the problems that have been identified in relation to genetic information and insurance,[5] or to rehearse all of the arguments that have been advanced for how we might best address them (Daniels et al. 1996; Human Genetics Advisory Commission 1997; various in McGleenan et al. 1999; various in Rothstein 2004). Nor will I provide a comprehensive comparative law analysis.[6] However, it will be instructive to the proposed endeavour to distil from the literature the principal problems raised by genetic testing in the context of insurance, and the benefits and pitfalls of actual regulatory approaches to it. Once this task has been completed, I shall take recourse to the fantasy scenario to defend the claims that none of the current approaches are palatable, and that they may in any case be futile. Here I shall reflect on how the core values of liberal political philosophy might guide the deliberations of legislators/regulators. It will be demonstrated that the only morally defensible, feasible approach to insurance, will be a state-run community-rated insurance scheme.

Insurance is a form of contract in which one person or group, the insuree, transfers certain risks or losses to the insurer for consideration. Underwriting, for which symmetry in information between insurer and insuree is needed, is the method used to classify people according to risk (Ewald 1999). Prior to the advent of genetic testing, individuals were already duty bound by insurance law to disclose their family history to insurers, for many policies. The argument now is over whether anyone who has already undergone genetic testing is likewise duty bound to disclose this health information to insurers at the time a contract is being formed. Certain kinds of genetic information are deemed by some insurers to be relevant to evaluation of the magnitude of the risk being transferred, for example, a test result indicating the individual carries the gene for Huntington's disease (UK Genetics and Insurance Committee 2000). In theory, this additional source of information permits insurers to classify individuals more accurately into various categories of risk or to assess risk premiums more accurately. Thus, relevant genetic test results, if disclosed, could enhance equity in insurance – the idea that people who have similar health or roughly the same life expectancy should pay equal premiums, and people who have worse health or a lower life expectancy should pay more.

The general ethical problem that genetic information poses for insurance is twofold: discrimination by insurers against individuals who possess what are known (statistically) to be offending genes, and adverse selection – this is not in fact much different from the twofold problem that other kinds of health information, actual or predictable, pose. I shall now address each prong of the general problem in turn.

Discrimination may occur in a number of guises. Those typically discussed include outright rejection of an application for coverage, exorbitant premiums for the same coverage and eligibility for only a basic coverage insurance package. Genetic information, like other health information, can be obtained by insurers through access to medical files – those of prospective insurees or of their family members. It can also be obtained from the people applying for insurance through medical history questionnaires; applicants are to disclose whatever a 'reasonable insurer' wants to know. Some genetic test results certainly qualify as information that reasonable insurers would want disclosed to them. One side-effect of the possibility of discrimination is that individuals may avoid being genetically tested, with grave health consequences, for fear of ramifications for their insurance. If I am not tested for fear of discrimination by insurers, I am denied detailed knowledge of my genetic predispositions and susceptibilities. This information might be valuable for decision-making about medical treatment and/or my lifestyle. Reluctance to be tested may not only impact adversely on my own health but also on that of family members or others.

Information gained from genetic testing might also be medically useful to people genetically related to me or, if I am planning a family, to my partner/spouse (Knoppers 1999).

Adverse selection, in broad terms, occurs when a negotiation between two parties who possess information asymmetrically affects the quality of the goods being traded. Usually this is because the party in possession of more information is able to negotiate an exchange that is relatively favorable to him or her. In the domain of insurance, a lack of symmetry in information between an insurer and insuree could reduce company profits, and also affect other insurees economically – for example, by a hike in insurance premiums across the board. Genetic testing arguably makes the incidence of adverse selection more likely and more severe, because the possibility exists for individuals to use even more types of information to gain significant benefit from both health and life insurance coverage/payouts.

4.1 Current Remedies – Benefits and Pitfalls

Different analysts use diverse criteria and nomenclature to categorize approaches that have been taken to address the two main problems outlined in brief immediately above.[7] I shall provide discussion of them under the umbrella terms of *self-regulation*, *moratoria*, *capped coverage* and *outright prohibition*. These four do not encompass all possible approaches. However, what I have chosen to omit and gloss over would not, were it included in discussion, undermine the main argument of this section of my chapter.

4.1.1 Self-regulation

It is has been argued that self-regulation by the insurance industry might adequately minimize discrimination, and that the incentive exists for the insurance industry to develop a code of practice regarding the use of genetic information that does so. This has occurred in some countries (Association of British Insurers 1997). To show why self-regulation (for example, limiting the types of genetic information sought from insurees) is unsatisfactory as a complete solution, I shall frame it as a collective action problem. The collective goal of insurance companies is to make as much money as possible. If it is factually the case that this goal would be *most* severely compromised by legislation that prohibited any access to genetic information, it would be rational for insurance companies to self-regulate with a view to averting this worst possible outcome. Self-regulation in the form proposed would result in sub-optimal profits but yield a better overall profit outcome than if prohibitive government legislation had been introduced.

When self-regulation is cast in this way, we can immediately see its chief potential weakness. Not everyone may abide by industry standards – there may be 'free riders', indeed, many companies may ultimately reject them (for example, when they realize they are losing competitiveness). Whether or not they do will greatly depend on the ability of the industry adequately to enforce its code of practice for genetics. As Otlowski points out, this ability, of course, may not in fact exist (Otlowski 2001). In addition, it is unclear that the insurance industry will be prepared to go far enough – their business, after all, is to turn a tidy profit. When we consider that the insuree is the weaker party to the insurance contract, a fact explicitly recognized by insurance and other kinds of law (Lemmens 2003), it seems that both adequate capacity to enforce regulation and adequately fair regulations are indicated. Moreover given the inbuilt unfairnesses of insurance, allowing even some exacerbation of them does not strike one as affording the best possible regulatory result. Finally, we claim here, as we also do in relation to other approaches to genetic testing and insurance, that it is conceptually incoherent to single out genetics as the sole subject of regulation. Bad genes are instances of brute luck just as are numerous non-genetic diseases.[8]

4.1.2 Moratoria

A number of countries have adopted a let's-bide-our-time approach to genetic testing and insurance, in the form of moratoria. At one point or another moratoria of varying descriptions have been imposed in Australia (Otlowski 2001), Canada (Knoppers et al. 2004), Finland (Knoppers et al. 2004), France,[9] Germany,[10] Greece, Ireland (Irish Insurance Federation 2001), the Netherlands (Lemmens 2003), New Zealand, South Africa, Sweden[11] Turkey, and the United Kingdom (UNESCO 1997). As noted by Knoppers and several other commentators, it is a common misunderstanding of moratoria that they *necessarily* involve complete non-usage of genetic information. Only approximately half of the countries just listed ever committed fully to the non-use of genetic test results, and even in these, modifications to the rule occurred.[12] In the other half, moratoria served to reinforce the status quo: insurers already request disclosure of certain genetic test results or they only apply to insurance contracts that do not exceed a certain amount of time, or they only deny insurers access to genetic information when policies are limited to relatively minimal coverage (Knoppers et al. 2004).

It is clear that moratoria are, in the main, regarded as a temporary measure only. This makes it the case that they are of cold comfort to individuals who are advised to obtain genetic testing. When a moratorium is lifted, any individual who had undergone testing when it was in place may now experience problems obtaining affordable insurance. However, the temporariness of moratoria may also be viewed as a positive feature in that they afford government and the insurance industry flexibility by giving both time to assess accurately the seriousness of the problems genetic testing raises. Government ideally wishes to avoid knee-jerk legislation and to encourage insurers to work with some conception of the public good in mind. Insurers ideally will want the opportunity to develop their own code of practice, presumably to engender public confidence, and to circumvent heavy-handed government interference in the industry.

An illustration of a moratorium working to the broad satisfaction of government and the insurance industry, and which highlights the points made by Knoppers et al. above, is the *Concordat and Moratorium on Genetics and Insurance* (HM Government and ABI 2005), which went into effect in March 2005. This document details agreement between government and the Association of British Insurers (ABI) on the use of predictive tests in insurance. It modifies and extends the 2001 moratorium to 2011. The *Concordat* endorses the commercial principle that, unless agreed to the contrary, insurance companies should have access to all relevant information to enable them to assess and price risk. The types of genetic tests that may be used by insurers must be approved by the government's Genetics and Insurance Committee (GAIC). Insurers will not use any information from predictive genetic test results to underwrite travel insurance, private medical insurance, or any other one-off or annual policy, or for long-term care policies. But, they are entitled to require test results for life insurance policies over £500 000, critical illness insurance policies over £300 000, and income protection insurance exceeding £30 000 per annum. The terms of this document show that the UK has adopted a mixed strategy to deal with the problems genetics raises for insurance: self-regulation, moratoria and capped coverage.

4.1.3 Capped coverage

The provision of capped insurance coverage is already common practice in the insurance industry where, for example, insurers underwriting very basic coverage packages have never required disclosure of genetic test results or of other health information. Capped coverage can and does take a variety of forms. The UK example above is one such form, and the example offered by the Netherlands is another. The Netherlands integrated certain of the provisions obtaining to capped coverage from its now defunct moratorium on the use of genetic testing by insurers, into national legislation: the Medical Examination Act 1998. This law guarantees the provision of capped insurance for both basic life and disability insurance.[13]

Is the provision of capped coverage an appealing solution to the two main problems – discrimination and adverse selection – genetic testing presents for insurance? People who have already undergone genetic testing or who have a poor family history are guaranteed access. When guaranteed access is not simply the product of self-regulation, but is enshrined in legislation it offers solid protection against flagrant discrimination. In addition, capped policies could help minimize adverse selection – people with highly risky health profiles may be less inclined to manipulate the terms of an insurance contract in their own favor by not disclosing test results, and people in high-risk categories pay equal amounts for the same coverage. The main pitfall of the approach is obvious. Even if capped coverage plans extend to the insurees' health profiles more generally, it is absolutely clear that some people will be far more advantaged than others in the insurance market. This is to complain about the morality of equity in insurance. As I argue later on below, this is a complaint that governments/regulators have good reason to take more seriously than they have to date.

4.1.4 Outright prohibition on the use of genetic information by insurers

Various pieces of international and national legislation have been introduced that prohibit outright the use of genetic information by insurers. For example, the Convention on Human Rights and Biomedicine (1997), makes it clear that predictive genetic testing should only be carried out for health or research purposes (Convention on Human Rights and Biomedicine 1997). The Convention is only legally binding if ratified. (At the time of writing 19 states have ratified it, 13 have read but not signed it.) Ratification binds states to bring existing national law into conformity with the provisions of the Convention.

Let us cast an eye at the wording of the outright prohibition against the use of genetic test results by insurers. In Chapter IV of the Convention we find:

> Article 11 (Non-discrimination): Any form of discrimination against a person on grounds of his or her genetic heritage is prohibited.
>
> Article 12 (Predictive genetic tests): Tests which are predictive of genetic diseases or which serve either to identify the subject as a carrier of a gene responsible for a disease or to detect a genetic predisposition or susceptibility to a disease may be performed only for health purposes or for scientific research linked to health purposes, and subject to appropriate genetic counseling.

In the Convention, which clearly proscribes the use of genetic information by insurers, there are no similar prohibitions of the use by insurers of other kinds of medical information, such as family history. In Lemmens's analysis of Belgian national law, introduced in 1992, according to which, medical tests may only be carried out for medical or research purposes, she finds the same kind of problem. Outright prohibition in the form of genetics-specific legislation, however well intentioned it is, is deeply flawed. One ironic effect is reverse discrimination. For example, in the Belgian context, those whose family health histories are grim cannot purchase coverage at lower rates because they cannot prove through genetic testing that they themselves do not possess the deleterious genes responsible for causing disease in the family. Relatedly, as Lemmens emphasizes, it is remiss of legislators not to be explicit about what they understand by 'genetic data' or by 'techniques of genetic research that are used to determine the future state of health' (Lemmens 2003). If, for example, a non-genetic test, which insurers could require insurees to disclose, reveals a genetic condition, then the outright prohibition on the use of genetic test results is effectively neutered – insurees are not protected. In addition, genetic-specific legislation does not account for the likely incidence of adverse selection. To date, the impact of outright prohibitions on the use by insurers of genetic test results in the 19 countries that have ratified the Convention is unknown. Finally, as we have been insisting all throughout our survey of regulatory approaches to genetic testing in insurance, it is plain that there is no morally meaningful difference between health information about genetic status and other kinds of health information, at the conceptual level. It is a particular disgrace that the human rights-oriented approaches overlook this fact.

4.2 Summary of Pitfalls

None of the current policy strategies regulating the use of genetic information by insurers are entirely satisfactory, though some are clearly more so than others. Self-regulation, whilst it should be encouraged, may insufficiently minimize discrimination and may, in any case, be unenforceable. Moratoria, although expedient, often tend to reinforce the status quo, and by design they offer no long-term solution. Outright prohibitions on the use of genetic information by insurers using national or human rights legislation have the effect of introducing a different unfairness, and cannot cogently justify a conceptual focus on only one form of brute luck. The lesson for the present is that an integrated strategy, some combination of the strategies surveyed above, is doubtless the only sensible way to proceed halfways effectively. However, even a composite of the above will not fully meet an adequate standard of fairness for a reason integral to the character of insurance itself – the norm of equity. I now turn to demonstrate that all of the difficulties I have highlighted also point to the view that private insurance may soon no longer be even feasible.

4.3 Genetic Testing and Insurance in the Hypothetical World

Recall that in the fantasy scenario sketched above, a battery of tests from birth will reveal, to a high degree of accuracy, an individual's risk of developing more single early or late onset genetic-based disorders, and more combinations of genes that predispose individuals to illness of various kinds than is currently possible. Recall also that a greater range of accurately predictive home test kits for a greater range of specific tests for a greater range of diseases/predispositions to disease, are on the market. And, that genetic profiling of newborns is current, and is reliably predictive of their future health status. (More refined testing, and methods of testing are most probably less fantasy than future reality.)

Such a scenario begs the obvious question of how regulators could possibly control the use of knowledge gained from home test kits by individuals applying for health or life insurance. It could of course ban such test kits. However, this move will be difficult to defend for several reasons. What strong reason can we adduce to constrain individual liberty in this way? As noted above, such reasons must be given before government can justifiably interfere with individual action. Is there any direct and immediate threat at stake to the critical interests of others? Clearly not. Disallowing test kits on the grounds that the information they might reveal could be misused in a contractual negotiation that has not yet occurred, takes us into the realm of science fiction where 'criminals' are hunted down *before* they break the law.[14] Perhaps the reason might instead be a paternalistic one: government might deem it necessary that counselling accompany any genetic test? This position is espoused in many mission statements, including the April 2005 Consultation Paper issued by the National Bioethics Advisory Committee in Singapore (Bioethics Advisory Committee 2005). But mandatory counselling does not accompany, for example, home pregnancy tests. And, of course, there is no mandatory counselling for individuals forced to confront a family medical history or for those who grapple with other kinds of medical information relevant to their own health status. What is the explanation for this attitudinal disparity on the part of government? Why should an individual be thought ill-equipped to deal with potentially bad news about genetic-based ill health and yet adequately equipped to cope with other kinds of health information? Finally, even if a government in the hypothetical world banned test genetic home test kits, these could doubtless be obtained in and from other countries. To put the point bluntly: if an individual can order the components required to make polio virus[15] in his or her garage, government is not going to be very successful at controlling the dissemination and use of home test kits.

Why is this relevant to the discussion of existing regulation? If these kits are faithfully predictive, and readily available, presumably many individuals will be tempted to use them to gain significant advantages in the insurance market. This will cause a number of critical problems. First, other things being equal, what relative amounts of coverage would be sought by citizens who gained genetic information using a home test kit? Those who test negative for the deleterious gene disorders would want relatively less coverage than people who test positive. This latter group would seek extensive coverage. As for the people who test positive for certain predispositions to disease, the amount of coverage desired would depend, for example, on the frequency of unavoidable exposure to, for example, disease-inducing environments.

Individuals who made use of home test kits might, of course, be duty bound to disclose such results at the time they were negotiating an insurance contract, but it is probable that a good many of us might shirk that duty unless that is, disclosure advanced our cause. This puts the insurance industry on the back foot. For fear of gross profit loss through adverse selection companies would, possibly for their very survival, have to insist that anyone who wanted to purchase insurance coverage, at least over a certain amount, would have to undergo genetic testing. According, however, to some current legislation, as we have seen they would not be able to do that. The regulatory position that genetic testing should only be conducted for clinical or research reasons seems utterly wild in the light of this extreme threat of adverse selection.

There is a further complication that is brought to the fore by the hypothetical scenario. The genetic profiling of newborns (UK Human Genetics Commission 2005) also compromises current insurance practices and suggests that current regulations are inadequate and misguided. Could a government comfortably deny parents the possibility of knowing the genetic status of their children when this might aid in combating actual disease or the prevention of it? It is hard to see how. Surely, possible abuse by parents of this information in the domain of insurance is not a strong enough reason to prevent newborn profiling. Yet, we can note that it would be very tempting to parents to capitalize on any information they did gain from profiling, for example, by taking out a hefty life or disability insurance policy on their child. Of course, insurers could require that tests results be disclosed. Profiling information could count as something a reasonable insurer might demand. However, newborns consent to nothing at all let alone genetic testing. Should an individual who was profiled by his or her parents be denied insurance or have to pay higher premiums upon reaching maturity, for genetic tests he or she had no control over? Think of all of those legislative provisions insisting that any genetic information that is used by insurers is the product of tests that the individual him or herself voluntarily consented to.

4.4 Mitigating Brute Luck Fairly

Throughout my analysis of current regulatory approaches I highlighted the fact that brute genetic luck is conceptually no different than many other kinds of brute health luck. All are traceable to unchosen differences between persons. Is it ever fair for the incidence of brute luck to make some worse off than others? According to the broad liberal conception I outlined above, it is the role of government to mitigate brute luck suffered by individual citizens, by obligating individuals to demonstrate concern for others by sharing in the costs of the bad genetic luck that they experience. This is because, other things being equal, good fortune is not permission to ignore the plight of others. It is morally incumbent on us as a community to mitigate the bad genetic luck that victimizes only some of us. Good health is an essential ingredient of all individuals' respective conceptions of what it means to lead a good life, however much the content of those conceptions differ in other respects. An additional reason to insist that people share in others' brute genetic luck relates to external environmental factors. If two people have a genetic predisposition to a certain illness, and only one of them is exposed to a condition-aggravating external environment (for example, sun or smog), they will be at different risk of developing disease (for example, skin cancer or asthma). The existence of health-endangering environments is, at least in part, the responsibility of society as a whole. Moreover the choices of where to work and reside are not, in many cases, 'free' ones – they are much circumscribed by the vagaries of the marketplace and social circumstance.

4.5 State-run Community-rated Insurance

In the light of the above discussion, and in the light of existing difficulties with regulatory approaches to genetic testing and insurance, we need a different focus to private insurance. It strikes me that the suggestion by thinkers such as Dworkin (2000), Chapters 2 and 9 and Sorrell (2002), that a state-run insurance system is the most fair and most feasible way forward, is accurate. Dworkin has modelled such an insurance scheme in the broader context of his theory of liberal egalitarianism. This view aims to provide a unified account of equality and responsibility in which equality depends not on some bare outcome but on a process of coordinated decisions taken by individuals who accept responsibility in the context of belonging to a community of equal concern (Dworkin 2000, Chapter 3, p. 122). On his account, a distribution of resources is fair if it reflects the different ways in which people have chosen to lead their lives, and when it does *not* reflect unchosen differences in their personal circumstances – where circumstances are defined as natural endowments and state of health. The hypothetical insurance scheme is designed to mitigate the adverse effects caused by differences traceable to these last. In the scheme, members of a community are required to imagine how much insurance protection against risk they would have purchased, if they had had the opportunity to do so on equal terms, and in view of what insurers would have charged for premiums in a competitive market.[16] Individuals who suffer from deficiencies in their mental and physical capacities, which can be traced to genetic make-up, non-genetic-based disease, or accident, are to be compensated monetarily according to the results of the scheme. Dworkin's insurance approach insists that a nation should spend what individuals, collectively, would spend under these hypothetical conditions, and also that it should use that aggregate expenditure to ensure that all have now what they would have then. Note that this sort of approach might serve as the basis for a kind of life insurance too.

Adverse selection of a kind is a possibility even with the prudent insurance approach. Questions will arise as to which health conditions individuals are responsible for. Should smokers, for example, be penalized in the form of higher premiums? This is a popular suggestion, and has some moral weight. It is unfair for some to be recipients of scarce medical resources or to receive monetary payouts for health deficits, if lifestyle choices that were avoidable lie at the root of the medical problem or insurance claim (Dworkin 2000, Chapters 13 and 8, resp. pp. 451–2 and 307–19). However, it merits emphasis that we need to be circumspect when evaluating which lifestyle choices justify higher premiums (de Beaufort 2002). For example, there may prove to be genes that predispose people to risky behavior and/or addiction. And, activities that we currently think of as being virtuous and noble, such as being incredibly hard-working, may aggravate the number one killer in the developing world – heart disease. What individuals can be fairly held responsible for is a decision society will have to face. Whatever is decided, needless to say, will have to meet a higher standard than 'moralism'.

4.6 Section Summary

It should now be plain why I judge current regulatory attempts to govern the use by insurers of genetics testing to be short-sighted, piecemeal, counterproductive, potentially damaging and unsustainable. A simple abstract thought experiment has helped us to see why. Moreover the solution I have presented, one grounded in the demand that one role of government is to mitigate the effects of brute luck, looks to be far more practical than current policy-makers and many commentators in the debate over genetic testing are apparently prepared to admit. In addition, this sort of insurance wholly removes the main worry that people might have for undergoing genetic testing, namely, the possibility of discrimination. The threat of adverse selection is present, as we have noted. But it is not fatal to the scheme we have envisaged in the way it is to current regulation when we project that onto a possible future.

5 THERAPEUTIC CLONING

Relative to genetic testing, therapeutic cloning is a technology very much in its infancy. Whereas we can plausibly predict, as we did above, that genetic testing methods and the scope of such tests will dramatically improve in the proximate future, a like projection in the case of therapeutic cloning is more of a stretch. This view notwithstanding, analysis of current regulation of therapeutic cloning does have something to gain from postulating a future world in which therapeutic cloning is in clinical application.

5.1 Defining Therapeutic Cloning

The technique of somatic cell nuclear transfer has a long experimental history, with pioneering work on it dating back to the early 1960s (Gurdon 1962). It can be used in reproduction (the first cloned mammal was Dolly the sheep [Wilmut et al. 1997], a total of seven additional species have been cloned in this way since 1997), and to produce embryos for use in the development of stem cell therapies – a concept termed 'therapeutic cloning'. SCNT involves removing the nucleus from a single cell (for example, a kidney cell) and transferring it to a donor egg that has had its own nucleus removed. The somatic nucleus in the reconstructed egg is reprogrammed by components within the egg, and aided by artificial stimuli, the composite begins to undergo embryogenesis. At five to seven days into its development the embryo is sacrificed and stem cells are harvested. The pluripotent stem cells that are derived from the embryo carry the nuclear genome of the patient. When they are used to treat that same patient they should not therefore cause any adverse immunological reaction of the kind that plagues organ transplant recipients in the absence of a good donor match. Therapeutic cloning could potentially be used to provide tailor-made cells, tissues and organs for patients with diseased or damaged tissue that requires either replacement or supplementation. The first derivation of ES-like cells from surplus IVF human embryos was achieved in 1998. There is currently only one published ES cell line created (in 2004) by using SCNT (Thomson et al. 1998).

5.2 The Problems: Scientific and Moral

There are two main sets of problems with therapeutic cloning: scientific and moral. The scientific challenges facing ES cell researchers are formidable (Hwang et al. 2004). Concerns abound regarding the safety and efficacy of clinical usage, as well as cost. It remains an open question as to whether other kinds of research will serve the needs of patients better. There is also a debate over whether it is even appropriate to conduct ES cell research in the light of the possibility of using adult stem cells in their stead. This type of claim bears directly on ethical debate over any form of ES cell research. I have argued elsewhere (Burley 2005, Chapter 27, first two sections) that the failure to appreciate important differences between adult cells, ES cells and embryonic germ (EG) stem cells has (1) skewed the focus of discussion towards the source of the cells, thereby ignoring (2) important clinical advantages that work with ES cells may bring about. There exist compelling scientific arguments for an in-tandem approach to stem cell research.

The focus of the moral debate has centred on the source of ES cells. Whereas few if any object to the use of adult stem cells in the laboratory or clinical settings, strong disagreement persists over whether the development of ES cell-derived therapeutics is morally justified. ES cell research of any kind necessarily involves the destruction of life at the blastocyst stage (five to seven days). In the one published ES cell line created using SCNT the number of embryos used was not inconsiderable (see below).

5.3 Legislative Measures

A picture of the attitudes of the world's nations to therapeutic cloning can be gained by examining the results of a recent vote in the United Nations (UN) on the United Nations Declaration on Human Cloning (United Nations Fifty-ninth General Assembly 2005). It called on Member States 'to adopt all measures necessary to prohibit all forms of human cloning inasmuch as they are incompatible with human dignity and the protection of human life'. Eighty-four countries voted in favour of the resolution.[17] Thirty-four countries voted against,[18] 37 abstained and 36 were absent.[19] Deep divisions between countries over the permissibility of creating embryos using SCNT are therefore evident. However, as with other like resolutions, this one is not legally binding on individual member states. Moreover, as with many such declarations, the provisions are vaguely worded and require interpretation. For example, 'inasmuch as they are incompatible with human dignity and the protection of human life' is wide open to interpretation. This view is bolstered by remarks reportedly made by a representative from the Republic of South Korea (which dissented) who said his country intended to continue to allow therapeutic cloning because it 'would reaffirm human dignity by relieving pain and suffering'. Also, we can note, without disrespect intended, that 'all measures necessary to prohibit all forms of human cloning' places women carrying identical twins in a seemingly precarious situation! In short, nothing much can be gleaned from the text of a declaration. In addition, the UN Declaration included reference to 'all forms of human cloning'. Because most people, let alone states, are opposed to human reproductive cloning, and because considerable confusion reigns as to the difference between the two applications of SCNT (a confusion that, indeed, has found its way into a national bill)[20] it is perhaps more instructive to examine individual states' national legislation/policy regarding therapeutic cloning, to paint a more accurate portrait of prevailing national perspectives.

Currently, nine countries (comprising approximately 2.5 billion of the world's population) permit the creation of embryos for ES cell research using SCNT: Belgium, China, India, Israel, Japan, Singapore, South Korea, Sweden and the UK.[21] A number of countries have adopted more restrictive policies that allow ES cell research, but only using embryos that were not created specifically for that purpose. The favoured alternative source is embryos superfluous to the needs of couples who have undergone in vitro fertilization (IVF). In this group are Australia, Canada, the Czech Republic, Denmark, Estonia, Finland, France, Greece, Hong Kong, Hungary, Iceland, Iran, Latvia, the Netherlands, New Zealand, Russia, Slovenia, South Africa, Spain, Switzerland and Taiwan.[22]

In the United States of America, a unique policy situation prevails. In 2001, federally funded stem cell research was permitted but only on a stipulated number of existing ES cell lines.[23] This policy, however, did not constrain ES cell research in the private sector, where it can be carried out on approved *and* non-approved lines, and does not rule out use of those created by SCNT. Also, the federal structure of the political system in the USA makes it possible for individual states to introduce state-specific laws, which may override federal policy. This has already occurred in one state, California.[24]

A singular approach to the regulation of ES cell research is that taken by Austria and Germany, both of which allow such research but only on lines created before a certain date. Austria disallows the creation of embryos for stem cell research by any method,[25] yet the importation of pre-existing stem cell lines is not prohibited.[26] In Germany, where the Embryo Protection Act 1990 explicitly prohibits creation and utilization of embryos for any purpose other than reproduction, and where the Stem Cell Act (Das Stammzellgesetz) prohibits the use of funds – public or private – to derive new human ES cell lines, the importation of and work on embryonic stem cells produced from supernumerary embryos (for which there was no payment) that predate 1 January 2002, is legal upon approval by a supervisory body. Germany has made it criminal for scientists to go abroad to work on newly created lines or to collaborate with scientists living elsewhere on lines that post-date the deadline. (Some interpret the law even to include communication about new lines by e-mail.)

What is so singular about these approaches is the implicit failure to recognize that if it is deemed wrong to use embryos in research, then it should not matter *when* they were created. The USA's stem cell register approach, federal policy, is also open to the same charge. Why should dates matter? The most obvious reply is that we cannot undo what has been done, but we can certainly prevent further abuses from occurring. The USA cannot convincingly maintain this position, of course, because it allows private sector research on newly created lines. Austria and Germany can. In the process, however, surely they are vulnerable to the charge that their limited endorsement of stem cell research is tantamount to an act of collusion with those who are producing new lines. After all, the same kind of research is conducted on the new lines as on the older ones. We can draw an analogy here with cases involving the mistreatment of human subjects in research. Suppose that useful data has been gathered from such research, and that Country X wants its researchers to use these data but not any further such data gathered by the same methods after a certain time. Suppose further that similarly useful data continue to be collected from similar kinds of experiments performed elsewhere. Is it remotely plausible for Country X to maintain that what its researchers do with the illicitly gathered first set of data is utterly morally unconnected to further sets of like data still being gathered? I find it difficult to see how this might be plausible.

5.4 The Current Ethical Debate Over the Ethics of Therapeutic Cloning

According to objectors, therapeutic cloning is wrong because it involves not only the destruction of human life but the creation of it for a purpose that involves this.[27] This additional feature is held to matter in the sense of making therapeutic cloning more unethical than research involving spare IVF embryos, for two reasons. First, the idea of creating a life and then destroying it is thought to be worse than destroying an early life that is already in existence but which is destined to perish in a freezer or be terminated at a specified cut-off date.[28] As I have argued elsewhere in relation to the sources of ES cells generally, this may be misguided for the reason that life created for use in research cannot sensibly be thought of as being a potential person because there was no intent ever that it should become one (Burley 2005, Chapter 27, first two sections). This argument relies on the notion that the concept of potentiality is necessarily a contingent one.

It would seem that, in the same vein, it is more of an ethical travesty to use IVF embryos in research because the intent did exist for them to become persons at some stage. The second reason given for objecting to therapeutic cloning relates to the facts that embryo reconstruction is still inefficient. The South Korean group reportedly used 242 oocytes and cumulus cells (from 16 unpaid donors), and achieved a cloning efficiency of 19 percent to 29 percent (from a scientific standpoint this in fact was viewed as encouraging because the figures are on par with those given for other species). The production of their one ES cell line (SCNT-hES-1) involved 30 blastocysts, an efficiency rate of just under 4 percent (Thomson et al. 1998). It merits stress, in the light of the argument concerning potentiality, that these embryos would never have come to exist were it not for the research, and the group's intention was never to create a person. Second, the efficiency objection is somewhat misleading because whether it is many embryos that are destroyed or only one, the one embryo will still matter morally very much to objectors. If it is wrong to destroy embryos then one should not endorse one approach to ES cell research that involves relatively fewer losses, and yet object to another approach, therapeutic cloning, because it involves relatively more. This, however, is precisely what countries that allow ES cell research but prohibit therapeutic cloning, are doing.

'Respect for human life' and 'respect for human dignity' are catchphrases in many international declarations and national legislation, not only that proscribe therapeutic cloning (as the afore-discussed UN Declaration appears to do), but also the use of embryos in research and/or ES cell research. Others, including myself, have complained in the past that the notion of human dignity lacks content such that its presence in legislation, policy or mission statements, without adequate qualification (never given), is vacuous, and therefore forms no kind of comprehensible objection, let alone a weak one. Yet, those who draft legislation persist in using it without qualification. There are of course some concepts that we all have difficulty enunciating clearly but which we have understanding of nevertheless. Moreover we can quite readily grasp in some cases the concept of an affront to human dignity – for example, the rape and slaughter of children by militia. But when we speak of an affront to human dignity in relation to embryos the concept is, for many people, murky, to say the very least. Quite why those who object to embryo research do not strain every nerve to make clear this application of the concept, is mystifying. Perhaps the thought is that 'human' when attached to 'dignity' is doing enough of the work.

It is held that respect for human life encompasses all lives possessing the properties of humanness, and from that property it follows that any human life (whether only a few days old or fully formed) merits the same sort of moral consideration – first and foremost the security of its very existence.

In this view, when the notion of 'humanness' is deployed, it usually references scientific fact (the entity possesses the requisite genetic make-up of a member of the species *Homo sapiens*) and a conception of ontology (in the case of certain religions this is ensoulment).

A simple thought experiment, not my own but which I have used frequently in the past, is whether in the face of some natural disaster, under conditions of scarcity, namely one rescue boat, we should save the 100 adults populating one island or the 100 embryos populating the other. Only one rescue is possible. Which island do we direct the rescue vessel to? Owing to the stress placed by objectors to therapeutic cloning/ES cell research on the moral status of early life, we surmise that they should say 'The boat should be dispatched to save the 100 embryos' or, perhaps, 'The captain should toss a coin'. However, even the staunchest of objectors to research on embryos would likely not give either answer.[29] It therefore strikes me that 'respect for human life' embodies something more than consideration of the properties of humanness – the notion that all human lives in virtue of being human are equal *qua* human.

Whatever our views about the importance of early life, our moral deliberations about what should be done in situations where resources are scarce or non-existent, and lives are threatened as a consequence, almost invariably involve consideration of respect for human life in the context of a broader conception than mere humanness. Although it would be fatuous to deny that objectors to ES cell research, in large measure, are motivated by moral concern for early human life, the view that falls far short of being remotely convincing is that the debate over stem cell research is only or even fundamentally about *human* status. In the rescue case just mentioned, when people answer that the adults should be rescued, they do not in so answering commit themselves to saying that early life does not matter at all. They might think it does very much, and so would grieve for the embryos that perish. More importantly, the nub of their view need not even be that early human life does not matter as much as adult human life, when considered strictly in terms of the property of humanness. What the nub of their view has to be, however, is that respect for human life is not simply about *humanness*.

A robust analysis of the morality of therapeutic cloning and ES cell research demands a less myopic focus. What is required is assessment of the permissibility of stem cell research in relation to the notion of how we might best respect human well-being as whole, not just early lives, and not simply on the basis that they are human. This broadening of focus is appropriate for two reasons: it serves to accommodate a wider array of morally relevant features of what is being argued about; and, as a consequence, it does not invite the same degree of polarity between views, hence bland policy/legislative outcomes that typically result when society is at odds on issues of deep-seated moral conviction. Let us now turn to consider this along with other issues in relation to our hypothetical world in which genetic technology, including SCNT, has advanced beyond present-day capacity.

5.5 Therapeutic Cloning in the Hypothetical World

Recall that in this world therapeutic cloning has the demonstrated potential to cure and to treat disease, and to repair damaged tissue, and is allowed in some nations. The question faced by legislators in the nations where it is not, is whether to allow the introduction of this technology. What, in this scenario, will be the character of the debate about therapeutic cloning? We are positing that an-already possible technique has become more efficient, and that ES cell therapies (for which there is proof of principle today) have been developed using SCNT. No one therefore will be arguing about safety or efficiency. But, they will surely still be arguing about cost, and about whether it is legitimate to use early human life in the service of human persons. It does not therefore seem as though we gain anything from the thought experiment in this case. However, there is one important feature that should not be overlooked. Consider that in this world there may be people for whom the medically best, possibly only, treatment may be an SCNT-derived stem cell therapy, and that it will not be available to them in their home country because of moral disagreement there over the use of early life in medical treatments. Should the prohibitive legislative position in their country be upheld? Suppose that it is. There will be no way to prevent the occurrence of medical tourism. Medical tourism is a feature of the world today, and is likely only to become more commonplace. But the mere incidence of medical tourism is no reason, in itself, for policy-makers to capitulate. Consider, for example, the position of the government of Ireland on abortion. When an Irish resident has an unwanted pregnancy and is intent on aborting, she can travel elsewhere to undergo a termination. Some have argued that this is damaging to the health of women. I raise the example, to bring forward a different point. As serious an occurrence as unwanted pregnancy may be, it is not as serious, in the majority of cases at least, as life-threatening or mobility-threatening illnesses.

Can we justifiably deny someone a life-saving treatment on the grounds that some or many members of society disagree on moral grounds, with the use of embryos in the development of that treatment? To what extent would people who hold those views be prepared to advance them at the expense of adult human life once the treatment was possible and available? It is one thing to deny a woman an abortion, it is quite another effectively to condemn very ill people to death. Now, let us recall the rescue example.

I claimed that individuals would choose to direct the boat to save the adults not the embryos. Would they answer in relation to the fantasy world scenario that we should allow the adults to perish in favour of preventing embryos from being created to help save them? It is possible that that they might still answer in the affirmative. This I think would be wrong. But suppose they did, do we have any other principled reply available?

Disagreement over when and why early life should be valued has what Ronald Dworkin (1993) has called, an 'essentially religious' character. By 'essentially religious' he does not mean that beliefs about the ways in which embryos matter morally are religious per se. Rather he thinks that such views are conceptually akin to beliefs about what the true religious path is, in the sense that both reference deep-seated moral convictions about *the sacred*. When confronted by differing attitudes of this kind, what is the appropriate response by government? In the case of religious belief, can government justifiably mandate that citizens worship one god and not another? One of the distinguishing and commendable features of modern societies is recognition by government that it may not rightly seek to enforce a faith or a style of religious worship on its citizens. Dworkin's suggestion, to repeat, is that views about the sanctity of early life do not differ conceptually from religious beliefs – both are about what is sacred. If this is right, then it follows we should not abandon government neutrality in the case of disputes over what, if anything, is sacred about early life. Recast in this way, disagreement over therapeutic cloning no longer is fundamentally about the point at which life assumes value or whether we are God's or the Blind Watchmaker's children. Instead, it is fundamentally about whether it is legitimate for government to legislate on such matters. People can reasonably disagree about which god merits the title, or if any god does for that matter. Likewise, as we have seen, people can reasonably disagree over what kind of status early human life has. The proper role of government, from the liberal account, is not to promote the rightness of one understanding of the sacred over another. Rather it is to create the conditions of freedom that will enable people holding divergent views on the issue, to act on these views. Reasonable disagreement can be had about the moral status of early human life in a way it cannot be cogently be had over the moral status of adult human persons. Contemporary regulation of therapeutic cloning reflects a lack of recognition of the fact that taking a policy stand on which view about the early status of human life is the right one, is a move as illicit as government telling us all which religion to embrace. It is not mindless or unreasonable to object to therapeutic cloning on the basis that embryos have value but it is unreasonable to enlist government to impose one's own conception of the sanctity of life on other citizens. No one who takes seriously the demand of religious tolerance in modern society would champion the abandonment of government neutrality on religious worship. Therefore, the idea that government should take a stand on other conceptually identical issues, must be vigorously resisted in the present and in the future.

6 CONCLUSION

Implicit in the notion of framework is the concomitant notion of structure. One desirable property of any structure intended to accommodate the vagaries of circumstance, is stability. If policies are the scaffolding of a regulatory framework, due attention to foundations must be paid. Yet it is precisely this emphasis on foundations that is lacking in virtually all policy processes/outcomes with regard to genetics. By introducing the two core values of liberalism, and by reflecting on the proper role of good liberal government to uphold them, a principled basis for regulation was laid. Against that background I argued that a community-rated, state-run insurance scheme that is targeted at mitigating brute luck with respect to health, genetic or otherwise, would do a far better job than existing regulatory efforts. This is not yet another impractical suggestion by a philosopher. On the contrary, as I demonstrated, if developments in genetics continue apace, current legislation governing genetics and insurance is likely to be self-defeating. Nothing I have said rules out private insurance as an additional insurance option. What my discussion makes clear is that it is misguided solely to concentrate policy discussion and legislative efforts on it.

In the second main section of my chapter, I argued, this time by reference to a widely (though inconsistently) held liberal conception of freedom, that it may not be legitimate for legislators to deny citizens access to therapeutic cloning. What the hypothetical world shows is that if and when SCNT-derived stem cell therapies have moved beyond the stage of proof of principle, moral decision-making on this issue assumes, by necessity, a different character. This is because we cannot easily ignore the well-being of all of the very ill/injured people in the world when we can actually help them with proven therapies. As indeed, we should be not be ignoring them today through restrictive regulation of research into such therapies.

NOTES

1. Silver (1997), for example, is utterly confident that even with respect to human reproductive cloning, it is a matter of when not if. As for anticipated advances in genetic testing, most of the predictions are sanguine. See the UK Human Genetics Commission

Report (2005), pp. 20–22, which states both that profiling is a technology that is feasible now, and that within the next 20 years it may be possible to produce a full genetic profile of a person for £1000 sterling.

2. Rawls (1971) is more prepared than Dworkin (2000), Chapters 2 and 3, to acknowledge up front the factual premises of human well-being. But Dworkin does say, in relation to equality, for example: 'Someone who is born with a serious handicap ["handicap" is used in a technical sense] faces his life with what we concede to be fewer resources, *just on that account*, than others do', p. 81 (my italics). And, in relation to liberty: '. . . we think lives led under circumstances of liberty are better lives *just for that reason*', p. 121 (my italics).
3. Numerous political philosophers hold this view, and take different approaches to dealing with the problem: all of the essays in Clayton and Williams (2000); Cohen (2004), pp. 3–29; Dworkin (2000), Chapters 1–7; Macleod (1998); Roemer (1994); Steiner (1994); van Parijs (2004), pp. 45–69; cf. Anderson (1999), pp. 287–337, who thinks that this focus on relative differences is misguided: 'What is the Point of Equality?'.
4. For the distinction between brute and option luck see Dworkin (2000), Chapters 2 and 9.
5. For an overview of issues connected to genetics and insurance, see ESHG, PPPC, January 2000.
6. Comparative law analyses on US statutes and the European Convention see: ESHG, PPPC, January (2000); on international see: Knoppers et al. (2004), pp. 173–94; Lawton (1997); on Canada and Europe see: Lemmens (2003), pp. 41–86; McGleenan and Wiesing (2000), pp. 367–85.
7. For two alternative categorizations and nomenclature see, Knoppers et al. (2004), pp. 173–94; Lemmens (2003).
8. There is an extensive literature specifically devoted to discussion of justice and brute luck. See for examples: Dworkin (2000), Chapter 2; van Parijs (2004), pp. 45–69.
9. 'Etude génétique des caractéristiques d'une personne: l'engagement des assureurs de la FFSA' (1994/1999), cited in Knoppers et al. (2004).
10. German Insurance Association, Voluntary Agreement on the Use of Genetic Testing for Insurance, cited in Knoppers et al. (2004).
11. Sweden, Agreement Between the Swedish State and the Swedish Insurance Federation concerning Genetic Testing (May 1999), cited in Knoppers et al. (2004); Laurie (2002), p. 149.
12. In Finland, France, Germany, the Netherlands, Sweden and the United Kingdom, modifications to moratoria occurred.
13. In the Netherlands coverage is capped for basic life insurance at 300 000 guilders and disability insurance at 60 000 guilders. The legislation there provides for adjustments every three years in reference to the cost-of-living index. Cited in Lemmens (2003).
14. This is the theme of the movie Minority Report, released in 2003.
15. Virologist Eckard Wimmer announced that his US team had built live polio virus from scratch using mail-order DNA segments and a viral genome map that is freely available on the internet (Cello et al. 2002).
16. For discussion of the weaknesses of the scheme see various in Symposium on Ronald Dworkin's Sovereign Virtue (2002).
17. Afghanistan, Albania, Andorra, Australia, Austria, Bahrain, Bangladesh, Belize, Benin, Bolivia, Bosnia and Herzegovina, Brunei Darussalam, Burundi, Chile, Comoros, Costa Rica, Côte d'Ivoire, Croatia, Democratic Republic of the Congo, Djibouti, Dominican Republic, Ecuador, El Salvador, Equatorial Guinea, Eritrea, Ethiopia, Georgia, Germany, Grenada, Guatemala, Guyana, Haiti, Honduras, Hungary, Iraq, Ireland, Italy, Kazakhstan, Kenya, Kuwait, Lesotho, Liberia, Liechtenstein, Madagascar, Malta, Marshall Islands, Mauritius, Mexico, Federated States of Micronesia, Monaco, Morocco, Nicaragua, Palau, Panama, Paraguay, Philippines, Poland, Portugal, Qatar, Rwanda, Saint Kitts and Nevis, Saint Lucia, Saint Vincent and the Grenadines, Samoa, San Marino, Sao Tome and Principe, Saudi Arabia, Sierra Leone, Slovakia, Slovenia, Solomon Islands, Sudan, Suriname, Switzerland, Tajikistan, the Former Yugoslav Republic of Macedonia, Timor-Leste, Trinidad and Tobago, Uganda, United Arab Emirates, United Republic of Tanzania, United States, Uzbekistan, Zambia (United Nations Fifty-ninth General Assembly 2005).

18. Belarus, Belgium, Brazil, Bulgaria, Cambodia, Canada, China, Cuba, Cyprus, Czech Republic, Democratic People's Republic of Korea, Denmark, Estonia, Finland, France, Gabon, Iceland, India, Jamaica, Japan, Lao People's Democratic Republic, Latvia, Lithuania, Luxembourg, Netherlands, New Zealand, Norway, Republic of Korea, Singapore, Spain, Sweden, Thailand, Tonga, United Kingdom.
19. *Abstain:* Algeria, Angola, Argentina, Azerbaijan, Bahamas, Barbados, Burkina Faso, Cameroon, Cape Verde, Colombia, Egypt, Indonesia, Iran, Israel, Jordan, Lebanon, Malaysia, Maldives, Mongolia, Myanmar, Namibia, Nepal, Oman, Pakistan, Republic of Moldova, Romania, Serbia and Montenegro, Somalia, South Africa, Sri Lanka, Syria, Tunisia, Turkey, Ukraine, Uruguay, Yemen, Zimbabwe.
    Absent: Antigua and Barbuda, Armenia, Bhutan, Botswana, Central African Republic, Chad, Congo, Dominica, Fiji, Gambia, Ghana, Greece, Guinea, Guinea-Bissau, Kiribati, Kyrgyzstan, Libya, Malawi, Mali, Mauritania, Mozambique, Nauru, Niger, Nigeria, Papua New Guinea, Peru, Russian Federation, Senegal, Seychelles, Swaziland, Togo, Turkmenistan, Tuvalu, Vanuatu, Venezuela, Vietnam (United Nations Fifty-ninth General Assembly 2005).
20. Republic of South Africa's National Health Bill (2003). See the provisions regarding stem cell research.
21. Belgium: Service Public Federal Santé Publique, Securité de la Chaine Alimentaire et Environnement 2003; China: Ministry of Health 2003; India: Indian Council of Medical Research (ICMR) 2004; Singapore: Singapore, Parliament of Singapore 2004; South Korea 2004; Sweden 2004; United Kingdom: Human Fertilisation and Embryology Act 1990; this was followed in 22 January 2001 by passage through the House of Lords, approved in December 2000 by the House of Commons, of a bill that permits the cloning of human embryos to derive stem cells. See also World Stem Cell Map 2005.
22. Australia 2002; Canada 2004; Denmark 1997; Estonia 1997; Finland 1999; Greece 2002; Hong Kong 2000; Iceland 1996; Netherlands 2000; New Zealand 2004; Republic of South Africa 2003; Spain 2003; Taiwan 2002. See also World Stem Cell Map 2005.
23. The NIH Human Embryonic Stem Cell Registry lists the derivations of stem cells that are eligible for federal funding.
24. For example, Proposition 71, approved by California voters on 2 November 2004, establishes a state constitutional right to pursue stem cell research, including through SCNT. Federal funding, however, is still only available for approved lines.
25. Article 9 of the Law on Medically Assisted Human Reproduction 1992, states that that fertilized human oocytes and cells derived therefore may not be used for purposes other than medically assisted reproduction; and any intervention into the germ-line is strictly prohibited.
26. Neither by the Austrian Reproductive Medicine Act (FMedG) nor by the Austrian Pharmaceuticals Act (AMG).
27. See Burley (2005). See also, Savulescu and Harris (2004) for an interesting discussion of beneficence and malificence in relation to the creation of embryos.
28. Storage dates differ from nation to nation and also within nations. In the UK, Section 14 of the Human Fertilisation and Embryology Act 1990 (Chapter 37) stipulates the statutory storage period in respect of embryos should not exceed five years.
29. I have never proposed a variation of this thought experiment that substitutes 100 primates for the 100 human adults, and asked whether they or the embryos should be saved. I would be most surprised, however, if the embryos were not favoured in this case either.

REFERENCES

Anderson, E.S. (1999), 'What is the Point of Equality?', *Ethics*, **109**, 287–337.

Association of British Insurers (1997), *Genetic Testing: ABI Code of Practice*, London: Association of British Insurers.

Australia (2002), Research Involving Human Embryos Act 2002.
Bioethics Advisory Committee (2005), *Ethical, Legal and Social Issues in Genetic Testing and Genetics Research: A Consultation Paper*, Singapore.
Burley, Justine C. (2005), 'Translational Medicine in Stem Cell Research', in Ariff Bongso and Eng Hin Lee (eds), *Stem Cells: From Bench to Bedside*, Singapore: World Scientific Publishing.
Canada (2004), Bill C-6: An Act Respecting Assisted Human Reproduction and Related Research, 27 March.
Cello, J., A.V. Paul and E. Wimmer (2002), 'Chemical Synthesis of Poliovirus cDNA: Generation of Infectious Virus in the Absence of Natural Template', *Science*, **297**, 1016–18.
Clayton, Matthew and Andrew Williams (eds) (2000), *The Ideal of Equality*, London: Macmillan Press.
Cohen, Gerald A. (2004), 'Expensive Taste Rides Again', in Justine C. Burley (ed.), *Ronald Dworkin and His Critics*, Oxford: Blackwell, pp. 3–29.
Convention on Human Rights and Biomedicine (1997), 'Convention for the Protection of Human Rights and Dignity of the Human Being with Regard to the Application of Biology and Medicine: Convention on Human Rights and Biomedicine 1997', Oviedo, 4 April.
Daniels, Norman, Donald Light and Ronald Caplan (1996), *Benchmarks of Fairness for Health Care Reform*, New York: Oxford University Press.
de Beaufort, Inez (2002), 'Justice, Genetics and Lifestyles', in Justine C. Burley and John Harris (eds), *A Companion to Genethics*, Oxford: Blackwell, pp. 325–33.
Denmark (1997), Act No. 460 on Medically Assisted Procreation in Connection with Medical Treatment, Diagnosis and Research, 10 June (amended 1 September 2003) and Act No. 503 on a Scientific, Ethical Committee System and the Handling of Biomedical Research Projects.
Dworkin, Ronald (1993), *Life's Dominion: An Argument about Abortion, Euthanasia, and Individual Freedom*, New York: Alfred Knopf.
Dworkin, Ronald (2000), *Sovereign Virtue: The Theory and Practice of Equality*, Cambridge, MA: Harvard University Press.
ESHG, PPPC (January 2000), European Society of Human Genetics, Public and Professional Policy Committee, 'Genetic Information and Testing in Insurance and Employment: Technical, Social and Ethical Issues, EUROGAPP Project 1999–2000', http://www.eshg.org/.
Estonia (1997), Embryo Protection and Artificial Fertilisation Act.
Ewald, Francois (1999),'Genetics, Insurance and Risk', in Tony McGleenan, Urban Wiesing and Francois Ewald (eds), *Genetics and Insurance*, Oxford: BIOS Scientific Publishers Ltd.
Finland (1999), Law 488/1999.
Greece (2002), Law No. 3089 on Medically Assisted Reproduction.
Gurdon, J.B. (1962), 'The Developmental Capacity of Nuclei Taken From Intestinal Epithelial Cells of Feeding Tadpoles', *Journal of Embryology and Experimental Morphology*, **10**, 662–740.
HM Government and ABI (2005), *Concordat and Moratorium on Genetics and Insurance 2005*, London: Department of Health.
Hong Kong (2000), Human Reproductive Technology Ordinance.
Human Fertilisation and Embryology Act (1990), www.hmso.gov.uk/acts/acts1990/Ukpga_19900037_en_2.htm.

Human Genetics Advisory Commission (1997), *The Implications of Genetic Testing for Insurance*, London: Department of Trade and Industry.
Hwang, W.S., Y.J. Ryu, J.H. Park, E.S. Park, E.G. Lee, J.M. Koo, H.Y. Jeon, B.C. Lee and S.K. Kang (2004), 'Reports – Developmental Biology'. Evidence of a Pluripotent Human Embryonic Stem Cell Line Derived from a Cloned Blastocyst', *Science*, **303**(5664), 1669–73.
Iceland (1996), Act on Artificial Fertilization, 55/1996.
Indian Council of Medical Research (ICMR) (2004), *Draft Guidelines for Stem Cell Research/Regulation in India*.
Irish Insurance Federation (2001), *Code of Practice on Genetic Testing*, Dublin: Irish Insurance Federation, www.iif.ie/media.htm.
Knoppers, Bartha M. (1999), 'Who Should Have Access to Genetic Information?', in Justine C. Burley (ed.), *The Genetic Revolution and Human Rights*, Oxford: Oxford University Press.
Knoppers, Bartha M., Béatrice Godard and Yann Joly (2004), 'Life Insurance and Genetics: A Comparative, International Overview', in Mark A. Rothstein (ed.), *Life Insurance: Medical Underwriting and Social Policy*, Boston: MIT Press.
Laurie, Graeme, T. (2002), *Genetic Privacy: A Challenge to Medico-legal Norms*, Cambridge: Cambridge University Press.
Lawton, A. (1997), 'Regulating Genetic Destiny: A Comparative Study of Legal Constraints in Europe and the United States', *Emory International Law Review*, **11**(2), 365–418.
Lemmens, T. (2003), 'Genetics and Insurance Discrimination: Comparative Legislative, Regulatory and Policy Developments and Canadian Options', *Health Law Journal*, 41–86.
Macleod, Colin M. (1998), *Liberalism, Justice and Markets*, Oxford: Clarendon Press.
McGleenan, T. and U. Wiesing (2000), 'Insurance and Genetics: European Policy Approaches', *European Journal of Health Law*, **7**, 367–85.
McGleenan, Tony, Urban Wiesing and Francois Ewald (eds) (1999), *Genetics and Insurance*, Oxford: BIOS Scientific Publishers Ltd.
Ministry of Health (2003), 'Ethical Principles on Assisted Reproductive Technologies for Human Beings and Human Sperm Bank', October.
Netherlands (2000), Bill Containing Rules Relating to the Use of Gametes and Embryos (Embryos Bill), Parliamentary Documents II, 2000/01, 27 423, Nos. 1–2.
New Zealand (2004), Human Assisted Reproductive Technology Bill (HART), October.
NIH Human Embryonic Stem Cell Registry, http://stemcells.nih.gov/research/registry/index.asp.
Otlowski, Margaret (2001), 'Implications of Genetic Testing for Australian Insurance Law and Practice' (*Occasional Paper No.1*), Tasmania: Law and Genetics Centre.
Rawls, John B. (1971), *A Theory of Justice*, Cambridge, MA: Belknap Press.
Republic of South Africa (2003), National Health Bill.
Roemer, John E. (1994), *Egalitarian Perspectives*, Cambridge, UK: Cambridge University Press.
Rothstein, Mark A. (ed.) (2004), *Life Insurance: Medical Underwriting and Social Policy*, Boston: MIT Press.
Savulescu, J. and J. Harris (2004), 'The Creation Lottery: Final Lessons from Natural Reproduction: Why Those Who Accept Natural Reproduction Should

Accept Cloning and Other Frankenstein Reproductive Technologies', *Cambridge Quarterly of Healthcare Ethics*, **13**(1), 90–95.

Service Public Federal Santé Publique, Securité de la Chaine Alimentaire et Environnement (2003), 'Loi Relative à la Recherche sur les Embryons In Vitro', 11 May.

Silver, Lee (1997), *Remaking Eden: How Genetic Engineering and Cloning Will Transform the American Family*, New York: Avon Books.

Singapore, Parliament of Singapore (2004), Human Cloning and Other Prohibited Practices Bill, bill no. 34/2004, first read 20 July 2004.

Sorrell, Tom (2002), 'The Insurance Market and Discriminatory Practices', in Justine C. Burley and John Harris (eds), *A Companion to Genethics*, Oxford: Blackwell, pp. 398–407.

South Korea (2004), *Life Ethics Law*, 29 January.

Spain (2003), Boletín Oficial del Estado, Ley 45/2003, de 21 de noviembre, por la que se modifica la Ley 35/1988, de 22 de noviembre.

Steiner, Hillel (1994), *An Essay on Rights*, Oxford: Blackwell.

Sweden (2004), Government Bill 2003/04: '148' and Committee on Health and Welfare Report 2004/05: 'SoU7', November.

Symposium on Ronald Dworkin's Sovereign Virtue (2002), *Ethics*, **113**(1).

Taiwan (2002), Taiwan Department of Health, *Ethical Regulations for Embryonic Stem Cell Research*.

Thomson, J.A., J. Itskovitz-Eldor, S.S. Shapiro, M.A. Waknitz, J.J. Swiergiel, V.S. Marshall and J.M. Jones (1998), 'Embryonic Stem Cell Lines Derived From Human Blastocysts', *Science*, **282**(5391), 1145–7.

UK Genetics and Insurance Committee (2000), *Decision of the Genetics and Insurance Committee Concerning the Application for Approval to Use Genetic Test Results for Life Insurance Risk Assessment in Huntington's Disease*, London: GAIC.

UK Human Genetics Commission (2005), 'Profiling the Newborn: a Prospective Gene Technology?', www.hgc.gov.uk, March.

UNESCO (1997), 'Universal Declaration on the Human Genome and Human Rights', adopted by the General Conference of UNESCO at its 29th session on 11 November 1997.

United Nations Fifty-ninth General Assembly (2005), Plenary, 82nd Meeting (AM), Press Release GA/10333, 3 March.

van Parijs, Philippe (2004), 'Equality of Resources versus Undominated Diversity', in Justine C. Burley (ed.), *Ronald Dworkin and His Critics*, Oxford: Blackwell, pp. 45–69.

Wilmut, I., A.E. Schnieke, J. McWhir, A.J. Kind and K.H.S. Campbell (1997), 'Viable Offspring Derived from Fetal and Adult Mammalian Cells', *Nature*, 385, 810–13.

World Stem Cell Map (2005), compiled by William Hoffman, http://mbbnet.umn.edu/scmap.html, 17 October.

PART III

GMOs and agricultural biotechnology: regulating risk

5. Constructing risks: GMOs, biosafety and environmental decision-making

Paul Street

1 INTRODUCTION

This chapter sets out to consider briefly the nature and construction of risks. After suggesting that risks are constructed and maintained within polymorphic social networks topologically extended within and constituent of TimeSpace, it will examine the implications that this entails for developing novel forms of political engagement in environmental decision-making. Focussing on environmental decision-making in relation to genetically modified crops, the chapter will suggest that there are three different categories of risks posed by these crops and will assess the extent to which international regulatory structures governing the trade in genetically modified organisms (GMOs) provide states with sufficient capacity to integrate concerns regarding these different categories of risk together with other relevant factors into their regulatory decision-making on the importation, deliberate release and commercial marketing of GMOs.

The chapter will argue that continuing attempts have been made to narrow the categories of risks associated with genetically modified crops in order to exclude a wide degree of choice for governments making decisions on the importation, deliberate release and commercial marketing of GMOs. This has been carried out through attempts to construct and maintain a narrow view of the risks associated with GM crops so that only those risks that can be justified by a scientifically sound risk assessment undertaken on a case-by-case basis are regarded as sufficiently valid to prevent their importation, deliberate release and commercial marketing. This process, which can be seen at work within discussions in both the Cartagena Protocol on Biosafety to the Convention on Biological Diversity (Cartagena Protocol 2000) and the World Trade Organization (WTO), is not only problematic insofar as it denies the social nature of risks but is also in contrast to a more generally recognized need to widen public participation in environmental decision-making that can be seen in international legal developments such as the Convention on Access to Information, Public Participation in Decision-making and Access to Justice in Environmental Matters (Aarhus Convention 1998).

2 CONSTRUCTING, NEGOTIATING AND STABILIZING RISK

The writings of Ulrich Beck from the early 1990s, and in particular the work contained within *Risk Society* (Beck 1992), include some of the most well known and influential reflections on the nature of risk in modern society. Beck asserts that the self-reflexive nature of high, or late, modernity has produced a disjuncture that creates and marks off 'risk society' from previous history. While Beck's portrayal of modernity produces a somewhat problematic central premise that constructs a global history where modernity, late modernity, high modernity, or even post-modernity exist as discrete periods in time, we can leave this aspect of his argument aside without detracting from his central observation that risks are social phenomena capable of being understood from an abundance of positions, all of which possess their own legitimacy.

Central to understanding Beck's thesis is the notion that 'risk society' is a society turned in on itself, or rather a form of rational scientific thought that is reflexively turned on the products of industrial scientific progress. As such, the reflexive processes that characterize modernity pose somewhat of a contradiction because, while society is driven forward by rationality, it is also its undoing. The very reflexive nature of modern thought results in the production of a society that is 'an issue and a problem to itself' (Beck 1996, p. 32).

Risk society is to some extent a world where the choices to be made are always between 'bads', between, for example, the fossil fuels that created the ozone hole and the ecological hazards of nuclear energy. The impacts of these 'bads' are always unevenly distributed, inequitably shared out between the inhabitants of the planet, creating new social fractures that perhaps unsurprisingly appear to fall along old fault lines. It is this concern with the distribution of risks that characterizes, for Beck, risk society as 'a phase of development of modern society in which the social political ecological and individual risks created by the momentum of innovation increasingly elude the control and protective institutions of industrial society' (Beck 1996, p. 27).

Building in some part on the work of Habermas, one outcome of modernity for Beck is that the legitimacy of political, scientific and other societal institutions are called into question. As such the ability to define the nature of risks extends beyond those institutions so that 'critique is democratized' (Beck 1996, p. 33). In other words risk society is not just about the proliferation of risks but the proliferation and generation of expertise about risks. As Beck comments, 'in risk issues, no one is an expert and everyone is an expert, because the experts presume what they are supposed to make possible and produce: cultural acceptance' (Beck 1994, p. 9)

Knowledge about risks needs to be regarded not only as socially constructed but also as heterogeneous, located within sets of social practices that mobilize human and non-humans. These narratives or 'modes of ordering' (Law 1994) about risks are generated and stabilized within polymorphic social networks topologically extended within TimeSpace. Risks arising from a given situation are simply not capable of being singularly defined or understood, they become political conflicts. The proliferation of expertise about risk means that it is not possible to make definitive claims about risk, indeed even the assertion that an institution like science speaks with a single voice, can be seen to conflict with the manner in which scientific work is undertaken. As Feyerabend (1975, 1987), Kuhn (1962) and all those involved in the sociology of science have repeatedly shown, science is not a monolithic body of knowledge. Indeed as Francis Crick's (1990) own description of his and Watson's groundbreaking work on the structure of DNA shows, serendipity, play and intuition can have just as important a role as rationality in scientific discovery.

Our understanding of science is therefore one that must by necessity regard all scientific work to be socially heterogeneous. The results of scientific research are neither based on a singular methodology nor rationally derived from a map of nature that is objectively knowable. For scientific work to succeed, however, collective action must be mobilized across differing social worlds with social actors finding enough agreement on methodological and substantive issues to produce relatively stable facts, all be it that these facts are both partial and relational. The model of science and scientific work that emerges is one that is both social and inherently reflexive. It is self-confrontational, producing an inherent pluralization of valid scientific knowledge.

Environmental risks are therefore sites of contestation, where multiple knowledge claims come to bear. None of these multiple claims can ever provide an impartial objective assessment of a risk as risks themselves do not exist in a natural world to be explained and understood by an objective body of scientific opinion. As Bruno Latour argues in relation to the ozone hole, it is 'too social and too narrated to be truly natural; the strategy of industrial firms and heads of state is too full of chemical reactions to be reduced to power and interest; the discourse of the ecosphere is too real and too social to boil down to meaning effects' (Latour 1993).

Much the same can be said in relation to GMOs and the potential risks they pose, for they also are far too social to be truly natural, the subject of continuing processes of territorialization, through the re-articulation and embedding of modes of ordering in TimeSpace. These strategies, in their many forms, are ongoing attempts to monopolize and define risks. These attempts are always fragile, never permanent and certainly not defendable by recourse to a definitive body of knowledge. GMOs then can be considered as boundary objects, sites of contestation upon which differing bodies of knowledge are brought to bear in continuing efforts to stabilize the facts about GMOs and their risks. In this, legal texts, such as the Cartagena Protocol, also play their part, articulating and embedding a narrative about the risks associated with GMOs that helps to both stabilize facts about GMOs as well as extending these in TimeSpace.

The de-monopolization of expertise in risk society results in a further consequence, a required 'reinvention of politics' (Beck 1994). As Beck argues 'the circle of groups to be allowed to participate can no longer be closed to considerations internal to specialists but must be opened up according to social standards of relevance' (Beck 1994, p. 29). As a result the structure of decision-making must be widened with all the participants aware that decisions in relation to risks have not already been made and remain merely to be implemented. The previous practice of experts and policy-makers formulating decisions behind closed doors needs to be transformed into a broad public dialogue involving a multitude of agents (Beck 1994, p. 29). The processes that create risk society, including the widening participation in decision-making, continue to deepen the crisis of legitimation faced by social institutions. However, we are left with no other choice. The only path open to us is to continue to broaden the public dialogue, to accept that decision-making must inevitably be widened in order to possess even the veneer of legitimacy, while at the same time resisting attempts to monopolize decision-making especially where these are based on claims to objectivity, rationality and universality.

3 RISKS, MORE RISKS AND GMOs

If we consider for a moment the potential risks posed by GM crops then it is possible to group these into three categories:

1. where a component of a GMO, or an associated required component, technology or practice, negatively interacts biologically with the environment;

2. where structural, or social changes resulting from changed farming practices have a negative effect on the environment;
3. where social and economic changes occur through the introduction of GMOs onto the market, which, although possibly not directly harmful to the environment, are nevertheless damaging to the interests of a government or its peoples.

Later in this chapter, I argue that there has been a concerted attempt during the development of the international regimes governing the safe transfer of GMOs between countries, not only to restrict the evaluation of the risks associated with GM crops solely to only one of these categories, the first, but also to limit the voices of those considered legitimate to speak on the risks associated with GM crops to a body of scientific knowledge portrayed as both objective and homogeneous. However, before moving on to consider these issues in a little more detail it is perhaps worth taking some time briefly to consider the current state of commercially grown GM crops.

In 2003, 67.7 million hectares of commercialized GM crops were grown in 18 countries (James 2003, p. 3). The vast majority of this, 63 per cent, was cultivated in the United States. However, four other countries, Argentina (13.9 million), Brazil (3.0 million), China (2.8 million) and South Africa (0.4 million), accounted for 30 per cent of global coverage between them (James 2003, p. 4). Four crops, maize, cotton, canola and soybean account for virtually all commercial GM production. In 2002, 51 per cent of total global soybean production was genetically modified; indeed in Argentina a massive 98 per cent of the 11.2 million hectares of soybean grown in 2001 was genetically modified. Although 18 countries now commercially grow GM crops there remains an animated international debate regarding their utility and safety. This ranges from a concern with the manner in which transgenic crops are being commercialized and developed to continuing concerns over the risks that GM crops pose whether economic, social, environmental, or relating to human and animal health.

As genetic engineering began to make the move from science fiction to science fact in the early 1970s, scientists found themselves disagreeing on the extent to which they should engage in a wider public debate on the potential risks of these new technologies. On the one hand James Watson, who with Francis Crick had first outlined the structure of DNA in 1953 (Watson and Crick 1953a, 1953b, 1953c) encouraged public engagement, writing that 'this is a matter far too important to be left solely in the hands of the scientific and medical communities' (Watson 1971). On the other hand Philip Abelson, the then editor of *Science*, voiced the fears of many in the scientific community on the consequences that such a debate could have, writing in an editorial that 'talk of the dire social implications of laboratory related genetic engineering is premature and unrealistic. It disturbs the public unnecessarily and could lead to harmful restrictions on all scientific research' (Abelson 1971).

When Herbert Boyer, Stanley Cohen, Robert Helling and Annie Chang successfully transferred genetic material from the toad *Xenopus laevis* into *Escherichia Coli* (*E. Coli*) in 1972, the scientific community became keen to show themselves as responsible, capable not only of assessing the risks involved in genetic engineering but in deciding the extent and nature of the regulatory structure that ought to govern them. As Susan Wright (1994) observes:

> it was in the interest of nearly all admitted to policy making arenas that they not jeopardize control of policy by dissension, apparent inaction, or acknowledging as central, dimensions of the issue that obviously transcended the expertise of scientists . . . the overwhelming tendency of the process . . . was to reduce the scope (and thereby the daunting complexity) of hazard evaluation by seeing broader social problems as falling outside the field of concern. Thus the genetic engineering problem was re-projected almost exclusively in terms of producing suitable safety precautions to contain hazards. In other words, it was reduced and redefined in terms that made it susceptible to a technical 'solution'. (Wright 1994, p. 158)

The intervening 30 years have in many ways seen little change to this genetic landscape. The life science community still disagrees about the risks involved but remains keen to assert itself as the prime and often only institution capable of assessing both the risks posed by GMOs and deciding the extent to which their development ought to be regulated. The existence of socioeconomic and ethical issues associated with GM technologies are often accepted while at the same time these are also frequently portrayed as resting outside the legitimate boundaries of risk assessment when it comes to determining the safety of GM technologies and products. The extent of the risks posed by GMOs have frequently been discussed elsewhere (Ho 1998; Lappe and Bailey 1999; Rifkin 1998; Rissler and Mellon 1996) and as such there is no need to catalogue them extensively, nor to draw any conclusions as to the substance of these risks, indeed the latter task is perhaps always doomed to failure insofar as no definitive calculations of risk can ever be made. However, it is worth considering the general nature of these risks and further exploring their categorization as drawn above.

Firstly, there are those risks posed by a component of a GMO, or an associated required component or technology, interacting biologically with the environment. In relation to plants for example, there are the potential pleiotropic[1] and epistatic[2] effects of transgenes in the original modified plants; the risks of cross-pollination with wild relatives; the risks of horizontal gene transfer[3] and then recombination, creating plants that contain the original transgene or transgenes; the possibility of 'super' pests and 'super' weeds developing increased immunity to the modified organism; and the well publicized prospect that a crop engineered predominately to protect against attack from, for example, the boll worm or European corn borer, such as Bt cotton or Bt corn, will also kill other insects such as butterflies or bees.

In the same category we can also include the wider biological risks posed to animals, including humans, through the possibility of vector recombination creating new strains of viruses; increases in antibiotic resistance resulting from horizontal gene transfer;[4] or increases in allergens resulting from the components of the modified organisms. It is also possible to extend this first category further in order to include some aspects of the practices associated with farming particular GM varieties, such as the nature of pesticide use or tilling, when compared with their equivalent non-GM varieties. Indeed the United Kingdom conducted a series of such assessments during the Farm-scale Evaluations of spring-sown genetically modified organisms (ACRE 2004; Royal Society 2003).

Following these farm-scale evaluations the UK government decided, on the advice of the UK Advisory Committee on Releases to the Environment (ACRE), not to grant permission for the commercialization of both GM beet and GM oil seed rape because, as ACRE reported, 'in contrast to conventionally managed beet and oil seed rape, these crops would most likely result in adverse effects on arable weed populations that in turn would be likely to affect organisms at higher trophic levels, such as farmland birds' (ACRE 2004). However, a risk assessment conducted at the same time on a different GM crop, a line of GM maize (T25) tolerant to the herbicide glufosinate ammonium, was considered to hold no adverse risks when compared with non-GM maize and, as a result, received the agreement of the UK government, in principle, to its commercial cultivation (see *Hansard* Vol. 418, Pt. 52, Col. 1381).

The second category of risks associated with GM crops still possess direct environmental impacts. However, these do not result from either components, or associated components of GMOs directly interacting biologically with the environment, but from structural social changes changing the nature of farming in a country. As such these risks are related to more general questions regarding the global liberalization of agricultural trade and its impact on the environment (Anderson 2002; Barraclough 2000; Dhar and Chaturvedi 1999; Gonzalez 2002; Murphy 1999, 2002; Ritchie and Dawkins 2000; Watkins 1995; Whalley 2004) than solely to the GM debate. So, for example changes to land use and/or the structure of farming in a country through the substitution of endogenous varieties of food crops, for GM food crops or cash crops such as GM cotton, pose a direct threat to agricultural biodiversity as well as a loss of knowledge about that diversity. Similarly a higher reliance generally on agro-chemicals than in existing farming practice might pose wider potentially detrimental effects for soil nutrition, water quality and human and animal health, than the direct substitution of one non-GM variety for another GM variety.

A third category of risks that potentially arise from GM crops do not have a direct impact on the environment but could result in other social and economic consequences for countries, for example, the risks to food security posed by a move from food crops for local consumption to GM crops produced for export; the financial risks to farmers and the agricultural sector from the higher prices farmers have to pay for the seeds; the risks to indigenous industries, such as local seed companies, or to existing export markets that might prefer to buy from states that did not grow GM crops and from which there was therefore no possibility of contamination between GM and non-GM crops.

If no conclusive statement can be made on the extent of any risks posed by GMOs it might be asked why it is worth drawing up this categorization of risks at all? The answer rests in attempting to understand and assess the extent to which international legal provisions provide states with the ability to take into account any of these considerations, or indeed other ethical and political considerations that cannot even be characterized as risks when regulating the importation, deliberate release and commercialization of GMOs. For the most part, proponents of GM technologies have argued that only the first category of risks is appropriate for consideration in regulating GMOs and even then the scope and assessment of risk should be narrowly defined. As the then US Agricultural Secretary, Dan Glickman, stated during a speech in 1997 to the International Grains Council 'as long as these products prove safe, we will not tolerate segregation. We will not be pushed into allowing political science to govern these decisions. The stakes for the world are simply too high' (USDA 1997), or, similarly in 1999 to the Food and Agriculture Organization (FAO), 'uniform and scientifically sound global standards under the Codex Alimentarius Commission and the International Plant Protection Convention must be developed. Only through a system rooted in science and untainted by ideology or profit-seeking can the best of biotechnology's potential be realized and its risks averted' (USDA 1999).

Leaving aside any cynicism regarding the extent to which Dan Glickman and the US Government were themselves untainted by ideology or profit-seeking, there are a number of points to be made at this juncture regarding Dan Glickman's claim that in relation to the regulation of GMOs we ought to be aiming for 'uniform and scientifically sound global standards'.

Proponents of GM tend to argue that the scientific assessment of risks can only be interpreted in one objective manner, an argument that as we saw above is at odds with the nature of scientific practice, for its very reflexive nature questions its own scientific propositions. Second, uniform global standards can never be achieved at least insofar as the assessment of risks in relation to GMOs and their deliberate release into the environment are concerned, because the potential interactions between GMOs and the environment change from location to location. As a United Nations Environment Programme (UNEP) report prepared for the Cartagena Protocol negotiations states, the risks posed by GMOs 'are often environment-dependent' so that 'an organism that is safe in one country is not necessarily safe in another country' (UNEP 1995, Para. 82). Third, why should 'political science' be excluded, why should governments be restricted only to a limited set of assessment criteria such as those in our first category of risks? Governments are ideological, they are elected to make ideological decisions, and ought to take into account in their decision-making the concerns of their citizens, whether these are scientific, socioeconomic, or ethical. As we shall see below, this epistemic conflict continues to play itself out in various fora not least in the continuing dispute in the WTO between the US, Argentina and Canada on the one hand and the EC on the other.

4 RISKING BIODIVERSITY

The Cartagena Protocol is intended to create a uniform international procedure for regulating the safe transfer of living modified organisms (LMOs) (Cartagena Protocol 2000).[5] At the centre of the Protocol lies an advanced informed agreement (AIA) procedure that will 'apply prior to the first intentional transboundary movement of living modified organisms for intentional introduction into the environment of the Party of import' (Article 7.1). Under the Protocol an exporter proposing to transfer an LMO into the territory of another for the first time must notify their intention to the relevant authority of the Party of import, thus triggering a decision-making procedure under Article 10. In accordance with Article 15 the intended Party of import will then need to conduct a risk assessment 'in a scientifically sound and transparent manner' (Annex III.3) in order 'to identify and evaluate the potential adverse effects of living modified organisms on the conservation and sustainable use of biological diversity in the likely potential receiving environment, taking also into account risks to human health' (Annex III.1). To comply with Article 10 of the Protocol the Party of import must inform the notifier of its decision within 270 days (Article 10.3).

Perhaps the most interesting aspects of the Protocol for our purpose are the tensions that exist within the text regarding the grounds upon which parties may make decisions on the importation of GMOs. These tensions are a direct result of the negotiating process and the need to accommodate a diverse range of positions held by the five main negotiating groups[6] if the Protocol was ever to be concluded (see Street 2001). One of the central issues from the early days of the negotiations was the extent to which the Protocol ought to deal with issues beyond those relating directly to the protection of biodiversity. The final text draws a distinction between different types of GMOs, so that, while those LMOs intended for food and feed come within the scope of the Protocol, although under a separate Article 11 procedure than those covered by the AIA procedure outlined above, the Protocol specifically excludes pharmaceuticals containing LMOs from its purview under Article 5.

However, disagreements relating to the scope of the Protocol did not extend only to whether it should deal with the risks of LMOs as they related to drugs, food and feed. They also existed over whether the Protocol ought to contain references to human health; whether it should allow parties to make decisions concerning the importation of LMOs on economic and social grounds; whether and to what extent the Protocol ought to include the ability of parties to apply a precautionary principle in their decision-making; and, the relationship the Protocol ought to have to international trade rules. Indeed such was the disagreement amongst the parties on these issues that at the Extraordinary Meeting in Cartagena of the Conference of the Parties to the Convention on Biological Diversity (ExCOP) at which the negotiations were due to be completed, 30 out of the 42 draft articles remained unresolved.

With the Cartagena meeting ending in failure and the Seattle WTO Ministerial upcoming, Canada (WT/GC/W/359), Japan (WT/GC/W/365) and the United States (WT/GC/W/288) made moves within the WTO to incorporate discussions on biotechnology into the ongoing negotiations on agriculture. However, following the inclusion within the revised Draft Ministerial Declaration of 19 October 1999 (JOB(99)/5868/Rev.1) of a bracketed reference to ensuring that trade in products of agricultural biotechnology be 'based on transparent, predictable and timely processes' (Para. 29) and, in Paragraph 71, to the establishment of a working party on biotechnology, a number of European environment ministers made it clear that they felt the Biosafety Protocol negotiations were the proper forum for dealing with issues relating to GMOs.

After a series of informal discussions, the ExCOP was resumed in Montreal, Canada a year after it had been suspended in Cartagena. The final text of the Protocol that emerged provides the parties, under

Article 26, with the ability to take into account 'socioeconomic considerations' (26.1) in their decisions on the importation of LMOs, while at the same time requiring that this be 'consistent with their international obligations' (26.1). A requirement, which for the reasons we shall see below, potentially runs counter to the very ability of countries to make decisions on socioeconomic grounds. The draft negotiating text from Cartagena included 18 references to socioeconomic conditions dotted throughout the text that had been contentious from the start.

In preparation for the Cartagena Protocol negotiations a panel of 15 experts chosen by the Convention on Biodiversity (CBD) secretariat was asked to draw up a document outlining the approaches to risk assessment and regulations already put in place by states in relation to GMOs. However, in its report (the Cairo Report UNEP/CBD/COP/2/7 Annex IV) the panel chose instead to pass its own opinions on the risks posed by biotechnology, the report stating that 'many of the initial concerns and fears [of biotechnology] have been allayed as experience and knowledge has accumulated' (Para. 26), it 'does not differ from other technologies' (Para. 68), nor 'imply any greater risk attached to products from modern biotechnologies vis à vis those arising from traditional ones' (Para. 23). Indeed in relation to risk assessment per se the panel considered that it should be 'restricted to the basis of objective parameters [as] socioeconomic aspects bring value judgements into the analysis which inevitably vary among countries and communities and from case to case depending on considerations other than the nature of the technologies themselves' (Para. 21).

The final text of the Protocol saw references to socioeconomic considerations removed from the general body of the text and placed in a single article and provides an unhappy and somewhat contradictory marriage between the ability of states to take decisions on the importation of LMOs on socioeconomic grounds under Article 26 and the requirement in Article 10 and Annex III that decisions be taken on the basis of a scientifically sound case-by-case risk assessment. A similar tension is also created by the inclusion of a precautionary principle within the AIA procedure and Annex III. The nature of the Protocol is such that parties are required to make their decision under Article 10 on the basis of a case-by-case scientifically sound risk assessment and only if that assessment is inconclusive is it then open to them to make a decision applying the precautionary principle contained in Article 10.6. In other words the precautionary principle contained within the Protocol is severely limited and certainly the ability of a party to the Protocol to maintain a blanket ban on GMOs is denied to them on a strict reading of the text. In addition although such a decision could potentially be justified on socioeconomic grounds under Article 26, the inclusion of the phrase 'consistent with their international obligations' is a direct reference, as with similar phrases in the text, to the WTO agreements. As such it is an attempt to limit the ability of states to justify decisions under Article 26 and certainly to limit the success of any possible arguments within the Dispute Settlement Body of the WTO that the Protocol represents some form of *lex specialis* or *lex posterior* and therefore takes precedence over the WTO agreements.

The relationship of the Cartagena Protocol to the WTO agreements was one of the central most contentious issues during the negotiations, which, despite their completion, remains essentially unresolved today. The Cairo Report had recognized that in relation to trade in biotechnology a 'lack of harmony among national regulatory systems may create non-tariff trade barriers in international trade' (UNEP/CBD/COP/2/7 Para. 82). While at an early Working Group meeting the Australian government voiced the concerns of itself, the US and Europe that the 'the outcome of negotiations on the Protocol would need to ensure that the instrument does not derogate from the provisions of the Agreement Establishing the World Trade Organization (WTO) or affect the rights and obligations of members of WTO' (UNEP/CBD/BSWG/2/2 p. 33).

Throughout the various different draft texts, diametrically opposed views of the relationship between the Biosafety Protocol and the WTO could be found (see further, Street 2001). On the one hand draft texts stated that the Protocol should be consistent with the non-discriminatory provisions of the WTO agreements, while on the other hand further options proposed that where there were inconsistencies between the Protocol and the Technical Barriers to Trade (TBT) or Sanitary and Phytosantiary measures (SPS) Agreements the Protocol should prevail with the parties waiving their right to bring a complaint in the WTO. By the Cartagena Draft text Article 34 stated that 'the provisions of this Protocol shall not affect the rights and obligations of any Party to this Protocol deriving from any existing international agreement to which it is also a Party *except where the exercise of those rights and obligations would cause serious damage or threat to biological diversity*' (emphasis added). The text of this article was a major sticking point for the Miami group at the Cartagena ExCop, who insisted that everything in emphasis, and that had in previous drafts been in brackets, be removed.

The final compromise reached in Montreal saw the removal of any articles from the draft text on non-discrimination (Article 22) and the relationship with other agreements (Article 31) replaced by a preambular text that emphasized that nothing should 'be interpreted as implying a change in the rights and obligations of a Party under any existing international agreements' while at the same time understanding that the clause 'is not intended to subordinate this Protocol to other international agreements'. In addition, it included as we have already seen in Article 26 the requirement that decisions taken on socioeconomic grounds be 'consistent with their international obligations'. As such the final text agreed at Montreal is a strange compromise that certainly reflects the decision-making procedures within the WTO and does nothing to resolve the tensions between the Protocol and the WTO agreements. While including both a precautionary principle and an article allowing parties to make decisions on socioeconomic grounds both of these are limited by the nature of the Protocol's text and the existence of the WTO to which we shall now turn.

5 ENVIRONMENTAL RISKS AND THE GATT

Since the inception of the WTO in 1995, and prior to that within the GATT, there have been a number of disputes involving measures applied by Members for ostensibly environmental purposes that have been challenged as inconsistent with either the GATT or other WTO agreements. The most pertinent to the present discussion is that bought by Argentina, Canada and the US against the EC in *European Communities – Measures Affecting the Approval and Marketing of Biotech Products* WT/DS292 (*EC Biotech Products*) that, at the time of writing, has still to be decided by the Dispute Panel. The WTO has no separate environmental agreement, although there is a reference to the objective of sustainable development in the Agreement Establishing the WTO. A number of WTO agreements hold implications for Members wishing to adopt measures aimed at protecting the environment, in particular the General Agreement on Tariffs and Trade (the GATT), The Sanitary and Phytosanitary (SPS) Agreement, the Agreement on Technical Barriers to Trade (TBT) and the Agreement on Agriculture. A Committee on Trade and Environment exists within the WTO whose remit includes a consideration of the relationship between the WTO agreements and multilateral environmental agreements (MEAs). However, the focus here will rest primarily on the GATT and the SPS agreement.

While the GATT covers measures relating to all goods produced and transferred across borders, the SPS Agreement covers only those products that are of themselves potentially injurious to human, animal, or plant health, such as say asbestos fibres, or that as a result of a method or process of production creates an intrinsically dangerous product, for example by spraying crops with chemicals, the residue of which changes the nature of the crop from something that was safe to eat into something that is potentially dangerous.

Under the GATT there should be no discrimination between 'like' products amongst members, so that a Member must extend the same favourable conditions that it grants to its most favoured nation equally to all other

Members of the WTO (Article I). Members are also required by virtue of Article III(4) not to discriminate between 'like' products produced internally and those imported from outside its territory by requiring that they are 'accorded treatment no less favourable than that accorded to like products of national origin in respect of all laws, regulations and requirements affecting their internal sale, offering for sale, purchase, transportation, distribution or use'.

Were the argument to be made that GM crops are 'like' non-GM crops we can see that Article III(4) of the GATT potentially limits the ability of Members to regulate GM crops in a manner different from that by which they regulate non-GM crops. However, under Article XX members can in particular circumstances make exceptions to the generality of GATT Article III. Although there is no explicit mention of the environment in Article XX, sub-section (b) allows members to make exceptions where 'necessary to protect human, animal or plant life or health' while subsection (g) allows them to do so where it relates 'to the conservation of exhaustible natural resources'.

Although criticism has been brought against the GATT for the manner in which disputes involving environmental measures have been dealt with, in the *US Shrimp/Turtle* case the appellate body of the WTO explicitly indicated not only that provided they satisfy the requirements of the chapeau and are applied in a manner that is neither arbitrary nor a disguised restriction on trade (*US Reformulated Gasoline* WT/DS2/9) that measures aimed at protecting the environment can be consistent with WTO obligations under Article XX (b) and (g) (see Para. 185), but also, that Members may be free to act multilaterally to protect the environment. As the appellate body stated in *US Shrimp/Turtle* (WT/DS58/AB/R) 'we have *not* decided that sovereign states should not act together bilaterally, plurilaterally or multilaterally, either within the WTO or in other international fora, to protect endangered species or to otherwise protect the environment. Clearly, they should and do' (Para. 186).

As one can see, in relation to GMOs this potentially opens up the possibility of a Member State defending a measure it might make to regulate GMOs on the grounds that it is implementing its obligations under an MEA, in this case the Biosafety Protocol. Indeed this argument has already been raised by the EC, although not heavily relied upon, in its first written submission to the Panel in the *EC Biotech Products* dispute when, after earlier drawing attention to the inclusion of a precautionary principle in the Protocol, which we shall consider further below, it states at Para. 112 'the European Communities considers that the interpretation of the relevant WTO agreements should be consistent with the requirements of the Protocol'.

Even more interesting for our discussion was the decision of the appellate body in the *EC Asbestos* (2001) dispute, which provides the possibility for the inclusion of a far wider consideration of the relevant characteristics of a product in assessing its 'likeness' to another. When asked to consider the extent to which asbestos fibres were 'like' other fibres capable of being physically substituted for asbestos, the appellate body stated that it was necessary to move beyond a simple test of substitutability. Thus the appellate body considered that in assessing likeness the Panel should have considered four criteria '(i) the physical properties of the products; (ii) the extent to which the products are capable of serving the same or similar end-uses; (iii) the extent to which consumers perceive and treat the products as alternative means of performing particular functions in order to satisfy a particular want or demand; (iv) the international classification of the products for tariff purposes' (*EC Asbestos*, Para. 101).

The most interesting aspect of this decision lies in the inclusion of consumer preferences as relevant in assessing the likeness of a product. Firstly, in relation to GMOs, consumers in Europe, for example, have shown themselves to be far more wary of GM crops and have stated a preference for non-GM products, as evidenced in the UK government-backed public GM Nation debate. Second, the decision of the appellate body potentially widens the grounds for defending a measure under Article XX beyond any scientific test seeking to correlate cause and effect, essentially bypassing the need to show that a measure relates to a proven environmental hazard through the ability to assert that consumers prefer products they perceive to be environmentally friendly and thus these are not 'like' non-environmentally friendly products for the purposes of Article III(4). As such this indirectly opens up a degree of public input into the decision-making process at least insofar as that public is vocal and significant enough to have some evidential validity. Thus while the GATT appears to provide possibilities for allowing regulations aimed at protecting the environment that are not based solely on scientific justifications, as we shall see, interpretation of the SPS Agreement (Agreement on the Application of Sanitary and Phytosanitary Measures 1994) by the appellate body appears to have been moving in completely the opposite direction.

6 THE SPS AGREEMENT, RISKS AND GMOs

The SPS Agreement is intended to allow Members to set their own levels of health protection while at the same time ensuring that sanitary and phytosanitary measures neither 'arbitrarily or unjustifiably discriminate between Members' (Article 2.2), nor act as disguised restrictions on international trade. Members are encouraged to harmonize their sanitary and phytosanitary provisions through the adoption of international standards set by, for example, the Codex Alimentarius Commission, the International Office of Epizootics and those organizations operating within the International Plant Protection Convention. However, this does not prevent Members from adopting their own higher standards (Article 2.2).

Under the SPS agreement measures must be applied 'only to the extent necessary to protect human, animal or plant life or health'. They must be 'based on scientific principles', and should not be maintained 'without sufficient scientific evidence' (Article 2.2). Although Article 5.7 of the SPS Agreement provides that where there is insufficient scientific evidence Members can take preliminary measures 'on the basis of available pertinent information' there is a requirement that those adopting SPS measures on this basis actively seek more information and objectively review the measures 'within a reasonable period' (Article 5.7). The agreement requires Members to base SPS measures on an appropriate risk assessment process (Article 5). These measures should bear a rational relationship to the risk assessment and 'not be more trade-restrictive than required to achieve their appropriate level of sanitary or phytosanitary protection' (Article 5.5).

It will be remembered that the during the Cartagena Protocol negotiations whether or not to include a precautionary principle as a key element of the text became one of the major sticking points. Similarly some of the parties to the Cartagena Protocol negotiations were keen to ensure that those parties who were also Members of the WTO would not derogate from their international trade obligations in respect to LMOs. In the end the final text of the Protocol included both a precautionary principle and a commitment by the parties to respect their international obligations. The result leads not only to an apparent conflict in the Protocol text but a conflict between the approach taken within the AIA procedure of the Protocol and that in the SPS Agreement and the manner in which it has been interpreted by the Dispute Settlement Body (DSB) of the WTO.

In the *EC Measures Concerning Meat And Meat Products* WT/DS26/AB/R (*EC Hormones*) dispute the appellate body discussed the nature and relevance of the precautionary principle to the SPS Agreement suggesting, in the process, that 'the precautionary principle indeed finds reflection in Article 5.7 of the SPS Agreement' (Para. 124). However, despite this the appellate body went on to conclude that 'the precautionary principle does not override the provisions of Articles 5.1 and 5.2 of the SPS Agreement' (Para. 125). Indeed in the later *Japan Apples* dispute the appellate body went somewhat further in limiting the ability of Members to take action in line with that which they might take in applying a precautionary principle to the risks posed by GMOs, when, in considering the ability of Members to take preliminary measures under Article 5.7 in 'cases where relevant scientific evidence is insufficient', it chose to prevent any wider interpretation of this provision to something resembling a precautionary principle by explicitly stating that 'insufficient' was not synonymous or interchangeable with 'uncertain'. Thus where there is sufficient evidence of a potential risk, including where that evidence is of a general nature, Members are required to carry out a case-by-case risk assessment, ensuring that their SPS measure bears a rational relationship to the outcome of the assessment.

What conclusions can we draw in relation to the ability of states to respond to the risks of GMO under both the SPS Agreement and the Cartagena Protocol? Firstly the approach adopted by the SPS Agreement is narrower than that provided for under both the GATT Article XX and the Cartagena Protocol. Where under the GATT, particularly following the approach adopted by the appellate body in *EC Asbestos*, it might be possible to justify regulations that restricted the importation of GM crops on the grounds that consumer preferences rendered these not 'like' their non-GM cousins, no such possibility would appear to exist under the SPS Agreement. Similarly where a decision to prevent the importation of an LMO under Article 26 of the Protocol can be made on socioeconomic grounds, the SPS Agreement requires not only that no measure can be taken without sufficient scientific evidence but that there also be a rational link between the identified risk and the measure.

Second, any Member of the WTO making a general decision, or even not making a decision, not to allow the importation of GMOs without the Member undertaking an individual risk assessment would appear to be in breach of the SPS Agreement. Where a Member might argue that it was putting into practice the precautionary principle contained in the Cartagena Protocol, the interpretation of the SPS Agreement by the appellate body in both the *EC Hormones* dispute and *Japan Apples* suggest that this would be received with little success. While the appellate body in *EC Hormones* suggested that measures may be taken on the basis of divergent, or minority scientific opinions 'from qualified and respected sources' (Para. 194) that evidence must not be of a general nature but 'sufficiently specific to the case at hand' (Para. 200). In addition given that there is sufficient scientific evidence to raise questions concerning the risks that GM crops might pose when released into the environment, Members would appear to have to carry out risk assessments on a case-by-case basis, ensuring that the measures they undertake are proportionate to the risk identified. Thus not only would Members appear to be unable to defend measures taken by them in relation to GMOs under Article 5.7 of the SPS Agreement, but the potential costs involved in conducting case-by-case risk assessments, especially for developing countries, are huge, although were this carried out under the mechanism contained with the Cartagena Protocol then Article 15.3 provides that states could recoup these costs from those seeking to import the GMOs. The cost of the farm-scale trials in the UK on just three varieties of three spring-sown crops was around £5 million. Unlike the risks posed to human health by food containing GM ingredients, the environment-specific content of the risk posed by GM crops means that the results of risk assessments are not generalizable. What is more, consideration needs to be given to the timescales of the risk assessment process. For we do not know how long it might take for the impact on the environment of GM crops, and their associated practices or technologies, to become apparent – something that is perhaps even more relevant to genetically modified trees given their lifespan.

7 CONCLUSION

The de facto moratorium on the commercialization of GMOs in the EC, which existed from 1998 until May 2004, arose on the one hand from the reluctance of individual EC Member States to allow the commercialization of GM before the existing regulatory framework had been enhanced, and on the other hand from the need of Member States to respond to internal public and consumer concerns regarding GM technologies and products. These concerns were not, and are not, always expressed solely in terms of a scientific assessment of risks posed by individual crop varieties or food products. In the UK anti-GM sentiments led, in September 2001, to the UK Agricultural and Environment Biotechnology Commission (AEBC) recommending that the UK government initiate a public debate to determine the nature and range of public views on genetic modification (GM) and the possible commercialization of GM crops. The results of the GM debate, and the methodology adopted, have been criticized for drawing on too narrow a sector of the public. However, an estimated 675 public meetings were held across the UK, with over 1200 additional letters or emails received by the debate, and in this regard at least can be contrasted favourably with the single conference and 316 contributions received by the EC Commission in its public consultation process *Towards a Strategic Vision of Life Sciences and Biotechnology* (SEC 2002).

The final version of *GM Nation? The Findings of a Public Debate* (UK Department of Trade and Industry/HMSO, 2003) was published in September 2003 and while recognizing its own limitations plainly states that:

> one point stands out. Among the participants in the debate there are *many more people who are cautious, suspicious or outrightly hostile about GM crops than*

> *there are supportive towards them*. Very few would support their early introduction on a commercial scale. There is a spectrum of preferred alternatives, ranging from an outright and indefinite ban to (more frequently) a period of delay to allow more information to be gathered about GM crops and their effects, and for tighter regulatory conditions to be imposed. (Para. 41 original emphasis)

Despite the results of the UK public debate on GMOs, on Tuesday, 9 March 2004, Margaret Beckett, the Secretary of State for the Environment, Food and Rural Affairs, made a statement to the House of Commons announcing the UK government's agreement in principle to the commercial cultivation of a line of GM maize. Only a passing comment was made regarding the regulatory framework for GM in the UK and certainly no mention was made of international legal obligations. However, given the positive farm-scale evaluation of the GM maize and ACRE's advice that permission should be granted for its commercialization, the UK government was placed in a difficult legal situation whatever its intentions may have been towards the findings of the public debate. For, as we have seen above, a decision not to allow commercialization of the GM maize in the circumstances would have been challengeable under WTO rules. Following the appellate body ruling in *EC Asbestos* an argument might be advanced that consumer preferences in the UK render GM maize unlike non-GM maize and that therefore a decision not to allow the commercialization of GM maize is defendable under the Article XX exceptions of the GATT. However, the limited possibilities that this argument might provide for taking account of public concerns in relation to environmental decision-making would not be available under the SPS Agreement given the manner in which the WTO dispute body have interpreted its provisions.

Beck's arguments on the nature of risk, together with the work of others such as Bruno Latour, John Law and Michelle Callon, allow us not only to challenge the narrow reductionist view of scientific work that is often mobilized in discussions of genetically modified crops, but also the manner in which decision-making regarding environmental risks ought to be undertaken. That our understanding of risk is constructed within social networks means that not only can risks never be described objectively, but that there is a resulting proliferation and demonopolization of legitimate positions on risk, leading to the conclusion that decision-making concerning these risks ought to be more inclusive. Indeed, such an approach has even been recognized by the Parties to the Aarhus Convention in its Guidelines on Access to Information, Public Participation and Access to Justice with respect to Genetically Modified Organisms and adopted by Decision I/4 at the first meeting of the Parties to the Convention held in Lucca, Italy, on 21–23 October 2002.

In many ways the UK can quite easily be seen to have developed a legal, regulatory and consultative framework that fulfils the requirements of the Aarhus Convention guidelines on GMOs both with respect to access to information and in relation to public participation, especially given that Paragraph 17 recommends only that due account should be taken of the outcome of any public participation with no decision required to be made in line with this, only that parties 'should, where appropriate and feasible, include an analysis of the comments and a description of the reasons for taking or not taking them into account in the (draft) decision'. However, Paragraph 21 provides more of a flavour of the spirit behind the guidelines when it recommends that, 'in order to improve public knowledge, public participation and awareness of activities involving GMOs, the public authorities are encouraged to explore other mechanisms and measures. Such mechanisms and measures could include consensus conferences, round-table discussions, stakeholder dialogues and citizens' juries on issues relating to, for example, the risk assessment and risk management of GMOs'.

It is at this point that we return again to Beck's notion of *Risk Society*. The need for more inclusive forms of environmental decision-making found in, for example, the Aarhus Convention, results at least in part from the recognition that not only are risks perceived differently by different sections of the public but that any resulting environmental harm is likely to be experienced asymmetrically. However, the very process of increased public participation in environmental decision-making subordinates and brings into question the legitimacy of political and scientific institutions. Increased public participation in environmental decision-making requires governments not merely to inform their citizens of the decisions they have made but to ensure that the public has sufficient information available for them to take an active and significant part in the decision-making process. If social institutions claim a monopoly on the decision-making process and governments merely adopt a facade of transparency while continuing to make decisions that do not reflect the outcomes of public participation, then ultimately the legitimacy not only of the decisions but the institutions themselves will be called into question. To some extent the crises of legitimacy that social institutions face will be the inevitable outcome of widening participation. However, the only acceptable option is to engage in increased public dialogue further enfranchising environmental decision-making while reinventing politics and law to further these ends as we go.

NOTES

1. Pleiotropy is the effect of the transgene on the functioning of other genes in the organism, thus affecting more traits than the one targeted in the modified organism.
2. Epistasis is the process through which one gene modifies the expression of another gene that is not an allele (alternative form of agene) of the first.
3. The transfer of genes from one organism to another by means other than reproduction (vertical gene transfer). A number of means of transferring and altering genetic material occur naturally through, for example, *viruses*, many of which cause diseases, or through parasitic genetic material such as *plasmids* and *transposons*.
4. In older GM varieties antibiotic-resistant genes were frequently used as markers, enabling genetic engineers to tag cells that took up the gene for the specific trait to be modified. Ciba-Geigy's (now Bayer's) transgenic maize, for example, contained a gene resistant to ampicillin, while Calgene's (now Monsanto's) modified FlavrSavr tomato contains a marker gene resistant to kanamycin. The fear is that these genes could transfer to humans through horizontal gene transfer bringing about increased resistance to antibiotics.
5. The Protocol uses the term living modified organism instead of genetically modified, transgenic or other term. This is partly because of earlier disagreements during the Biodiversity Convention negotiations concerning a suitable definition of genetic or/and biological modification.
6. 'The Miami Group' (Argentina, Australia, Canada, Chile, USA and Uruguay); 'The Like Minded Group' in essence G77 and China; The EU; The Central and East European Countries (CEE); and 'The Compromise Group' (Japan, Mexico, Norway, The Republic of Korea and Switzerland). For different perspectives on the negotiations from these groups see Bail et al. (2002).

REFERENCES

Aarhus Convention (Convention on Access to Information, Public Participation in Decision-making and Access to Justice in Environmental Matters) (1998), Aarhus (DK), 25 June 1998 (entered into force 30 October 2000) *International Legal Materials*, **38**(3), 517.

Abelson, P. (1971), 'Editorial: Anxiety About Genetic Engineering', *Science*, **173**(3994).

ACRE (Advisory Committee on Releases to the Environment) (2004), *Advice on the Implications of the Farm-scale Evaluations of Genetically Modified Herbicide-tolerant Crops*.

Agreement on the Application of Sanitary and Phytosanitary Measures (SPS Agreement) (1994), *International Legal Materials*, **33**, 1125–53.

Anderson, K. (2002), 'Developing Country Interests in WTO-induced Agricultural Trade Reform', in R. Adhikari and P. Athukorala (eds), *Developing Countries in the World Trading System*, Cheltenham, UK and Northampton, MA, USA: Edward Elgar.

Bail, C., R. Falkner and H. Marquard (eds) (2002), *The Cartegena Protocol on Biosafety: Reconciling Trade in Biotechnology with Environment & Development*, London: Earthscan/Royal Institute of International Affairs.

Barraclough, S. (2000), *Meanings of Sustainable Agriculture: Some Issues for the South*, Geneva: The South Centre.

Beck, U. (1992), *Risk Society: Towards a New Modernity*, London: Sage Publications.

Beck, U. (1994), 'The Reinvention of Politics: Towards a Theory of Reflexive Modernization', in U. Beck, A. Giddens and S. Lash (eds), *Reflexive Modernization: Politics, Tradition and Aesthetics in the Modern Social Order*, Cambridge: Polity Press, pp. 1–55.

Beck, U. (1996), 'Risk Society and the Provident State', in S. Lash, B. Szerszynski and B. Wynne (eds), *Risk, Environment & Modernity: Towards A New Ecology*, London: Sage Publications, pp. 27–43.

Cartagena Protocol on Biosafety (2000), Montreal (Canada), 29 January 2000 (entered into force 11 September 2003) *International Legal Materials*, **39**(5), 1027.

Crick, F.H.C. (1990), *What Mad Pursuit: A Personal View of Scientific Discovery*, Harmondsworth: Penguin.

Dhar, B. and S. Chaturvedi (1999), *WTO Agreements and Agricultural Sector: Implications and Options for India*, New Delhi: Research and Information System for the Non-aligned and Other Developing Countries.

EC Asbestos case (2001), *European Communities – Measures Affecting Asbestos and Asbestos-containing Products*, WTO Appellate Body Report and Panel Report, adopted on 5 April 2001.

Feyerabend, P. (1975), *Against Method: Outline of an Anarchistic Theory of Knowledge*, London: NLB.

Feyerabend, P. (1987), *Farewell to Reason*, London: Verso.

Gonzalez, C. (2002), 'Institutionalizing Inequality: The WTO Agreement on Agriculture, Food Security, and Developing Countries', *Columbia Journal of Environmental Law*, **27**, 433–90.

Ho, M. (1998), *Genetic Engineering Dream or Nightmare? The Brave New World of Bad Science and Big Business*, Bath: Gateway Books.

James, C. (2003), *Preview: Global Status of Commercialized Transgenic Crops: 2000*, ISAAA Briefs No. 30, Ithaca, NY: ISAAA.

Kuhn, T. (1962), *The Structure of Scientific Revolutions*, Chicago: Chicago University Press.

Lappe, M. and B. Bailey (1999), *Against The Grain: The Genetic Transformation of Global Agriculture*, London: Earthscan.

Latour, B. (1993), *We Have Never Been Modern*, London: Harvester Wheatsheaf.

Law, J. (1994), *Organizing Modernity*, Oxford: Blackwell.

Murphy, S. (1999), 'Market Power in Agricultural Markets: Some Issues for Developing Countries', *T.R.A.D.E. Working Papers No. 6*, Geneva: The South Centre.

Murphy, S. (2002), 'Structural Distortions in World Agricultural Markets', *Columbia Journal of Environmental Law*, **27**, 605–11.

Rifkin, J. (1998), *The Biotech Century*, New York: Jeremy P. Tarcher/Putnum.

Rissler, J. and M. Mellon (1996), *The Ecological Risks of Engineered Crops*, London: The MIT Press.

Ritchie, M. and K. Dawkins (2000), 'WTO Food and Agriculture Rules: Sustainable Agriculture and the Human Right to Food', *Minnesota Journal of Global Trade*, **9**.

Royal Society (2003), 'The Farm-Scale Evaluations of Spring-sown Genetically Modified Crops', *Philosophical Transactions, The Royal Society of London, Series B*, **358**(1439), 1777–913.

SEC (2002), *Towards a Strategic Vision of Life Sciences and Biotechnology*, 630.

Street, P. (2001), 'Trading in Risk: The Biosafety Protocol, Genetically Modified Organisms and the World Trade Organization', *Environmental Law Review*, **3**(4), 247–63.

UK Department of Trade and Industry/HMSO (2003), *GM Nation? The Findings of a Public Debate*, London: Department of Trade and Industry.

UNEP (United Nations Environment Programme) (1995), *Report of the Open Ended Ad Hoc Group of Experts on Biosafety* (Cairo Report) UNEP/CBD/COP/2/7.

USDA (United States Department of Agriculture) (1997), 'Remarks of Secretary Dan Glickman to the International Grains Council, London, England', USDA Press Release No. 0196.97, 19 June.

USDA (United States Department of Agriculture) (1999), 'Remarks as Prepared for Delivery by Secretary of Agriculture Dan Glickman to the U.N. Food and Agriculture Organization 30th Conference, Rome, Italy', USDA Press Release No. 0453.99, 13 November.

Watkins, K. (1995), *Agricultural Trade and Food Security*, Quezon City, Philippines: Oxfam.

Watson, J. and F. Crick (1953a), 'Genetical Implications of the Structure of Deoxyribonucleic Acid', *Nature*, **172**, 964–7.

Watson, J.D. (1971), 'Moving Towards the Clonal Man', *Atlantic*, 53.

Watson, J. and F. Crick (1953b), 'Molecular Structure of Nucleic Acid; A Structure for Deoxyribose Nucleic Acid', *Nature*, **171**, 737–8.

Watson, J. and F. Crick (1953c), 'The Structure of DNA', XVIII, *Cold Spring Harbour Symposia on Quantitive Biology*, 123–31.

Whalley, J. (2004), 'Environmental Considerations in Agricultural Negotiations in the New WTO Round', in M.D. Ingco and Winters (eds), *Agriculture and the New Trade Agenda: Creating a Global Trading Environment for Development*, Cambridge: Cambridge University Press, pp. 386–400.

Wright, S. (1994), *Molecular Politics: Developing American and British Regulatory Policy for Genetic Engineering, 1972–1982*, Chicago: Chicago University Press.

6. Legal framework and political strategy in dealing with the risks of new technology: the two faces of the precautionary principle

Wolfgang van den Daele

1 INTRODUCTION: ESCALATING RISK PERCEPTIONS

Conflicts over the risks of new technologies are endemic in modern societies. This does not, however, prove that we are living in a 'risk society' in which the risks of technology are out of control (Beck 1992). Independent of whether the risks grow, perceptions of risk are growing because the development of technology is out of control.

One of the problems with regulating modern technology in general and biotechnology in particular, is that regulation does not reflect the issues that are at stake in the political battles waged over these technologies. What is at issue is the structure and the process of modernization and, more precisely, the democratic control over the forces of social change. These issues are not on the regulatory agenda, however. Instead, regulation is predominantly confined to the containment of risks. The procedures designed to resolve conflicts over new technology hence fail to address the major themes that drive these conflicts.

Modern societies are structurally biased in favour of technological innovation for at least three reasons:

1. they have institutionalized a type of science that pursues objective knowledge and therefore accumulates truths that can easily be converted into technology;
2. they have adopted capitalist market economies that are based on imperatives of growth and innovation and provide, so to speak, a permanent societal opportunity structure for exploiting technological options; and
3. they acknowledge a system of rights that legitimizes the production of and access to new technology, both at institutional and individual levels.

As a result, a liberal regime of innovation has been adopted that limits the mandate of the state to control the social dynamics of technology through political decision-making. Restrictions of new technologies must be balanced with guaranteed freedoms. Justifications must be advanced for restricting new technologies, not for introducing them. Evidently, safeguarding important private and public goods, such as health and environmental safety, constitute sound reasons for restricting new technology. However, such restrictions do not consider the broader social and cultural impacts of technology. Neither have they served to bring social change resulting from technological innovation under democratic control. In view of the consolidation of free trade policies and the transnational positioning of major social players, such control now seems more remote than ever.

On the other hand, it is precisely the quest for democratic control of innovation that is at issue in most conflicts over modern technology. As governments and parliaments are unable or unwilling to put a revision of the basic features of the liberal regime of innovation on their agenda, conflicts over risks present themselves as a substitute arena. Escalating risk perceptions serve as a strategy to maximize political control over technology within the liberal regime of innovation.

There is ample opportunity for such a strategy. Risk is a psychological and social construct. Its perception is fluid, varies according to personal experience and interest, and depends on organizational culture and trust in institutions (Slovic 2000). The perception of risk can be manipulated through tactics of framing and rhetoric, and is hence not only a resource but to a certain extent also a product of political processes. By stipulating that uncertainty about the existence of suspected risks does not exclude restrictive action to control such risks, the rise of the precautionary principle (PP) on both national and international levels lends additional support to this strategy.[1] Critics of modern technology presume that the PP will widen the policy windows for democratic control of the dynamics of technology (O'Riordan and Cameron 1994; Tait/Levidow 1992; Wynne 1992). Accordingly, escalating the awareness of uncertainty is implied in the strategy of escalating risk perceptions.

However, this strategy has severe limits. Risk perception does not prove real risk. Risk perceptions are socially constructed, but they may also be 'fabricated', based on suggestive rhetoric or prejudice, induced through distortion of facts or other forms of manipulation. While it may be true that all reality is perceived reality, the reverse does not follow, and not all perception is reality. Even with a rigid constructivist epistemology, it is still possible and necessary to distinguish construction from reality. Thus, perceptions of risk may be genuine, but this does not prove that risks are genuine. Risk perceptions indicate fear, and whether such fears are grounded in fact is a different matter. Ascertaining the factual basis of such fears appears inescapable if restrictions of innovation are to be based on risk perception. This is true, at least, when the PP operates as a legal rule to be applied by courts in dealing with arguments revolving around risk and uncertainty.

2 PRECAUTION AS A LEGAL PRINCIPLE

Legal rules designed to deal with risks and uncertainties will be illustrated here with special reference to German law, which probably has the longest history in Europe in explicitly applying the PP for the control of new technology. We start with a recent case decided by the Administrative Courts (Verwaltungsgerichtshof of Baden-Württemberg 2002 and Bundesverwaltungsgericht 2003). In this case, the plaintiff argued that the authorization of a nearby production site was unlawful because it allowed emission of nano particles, which could have adverse health effects. The Court rejected the argument, giving as the main reason that the admissible emissions from the production site could not exceed 1 per cent of the threshold value set by the law for the saturation of air with nano particles from diesel, mainly stemming from traffic. The additional load from nano production was considered negligible. Table 6.1 summarizes the arguments of risk and uncertainty advanced by the plaintiff and the responses of the Court. The arguments of the plaintiff refer both to the reality of risks and to their evaluation, and they aptly demonstrate the escalation of risk perceptions and awareness of uncertainty that have become standard in all conflicts over new technology (Verwaltungsgerichtshof of Baden-Württemberg 2002).[2]

The responses of the Court display a logic of dealing with the risks and uncertainties of nano particles that may be summarized in four points:

1. Risk assessment precedes risk evaluation. It must first be established whether adverse effects are possible or likely to occur (does the risk exist?). As a second step it must be determined whether in light of such effects the technology should or should not be allowed (is the risk acceptable?).
2. Causal analysis is needed to establish the probability or possibility of adverse effects. Such an analysis is a cognitive process, not a normative one. Wherever such an analysis cannot be provided through recourse to everyday experience, it must be based on available scientific expertise and knowledge.
3. The burden of proof rests with those demanding the restriction of the technology (who must substantiate the possibility of adverse effects).

Table 6.1 Discourse over the risks of nano technology

Arguments of Plaintiff	Responses of the Court
Nano particles can have adverse health effects	On the basis of the scientific expertise consulted by the Court such effects are considered possible when the particles become incorporated through the respiratory tract
Nano particles could also be carcinogenic	Scientific expertise does not exclude that such effects may be possible, but considers them unlikely because health risks from nano particles are correlated with the number of particles, and not (as in the case cancer risks) with their mass [the value for diesel particles takes carcinogenicity into account]
The threshold value for diesel particles has no relevance for nano particles because it relates to the mass of the particles, not to their number	Available scientific expertise says that the value can be used as a basis for evaluating nano particles; the different baselines can be matched through calculation
Nano particles could have adverse health effects beyond carcinogenicity; they might, e.g., reach the brain or lungs by passing through cell membranes	The expert consulted by the Court could not confirm these effects. Rather, such fears refer to areas where the scientific data are contradictory or inconclusive
[The possibility of health effects beyond its carcinogenetic properties must be taken seriously]	The scientific expert explained that insufficient data are available at present to assign a degree of probability to such fears. The legal concept of danger does, however, require some probability that harm will occur
The threshold value for diesel particles does not account for the possibility that nano particles may have various degrees of substance-specific toxicity	According to a worst-case calculation proposed by the scientific expert, one is erring on the side of caution if the load of particles is 100 times lower than the value for diesel particles. The expected emissions from the contested production site are within this limit
[We lack the knowledge to assess the risks of nano particles; it cannot even be excluded that the risks are as high as with nuclear energy]	Existing knowledge says that the risks from nano particles cannot be compared with the risks from nuclear energy. Sixty to 90% of all nano particles to which people are exposed stem from automobile traffic
The history of toxicology proves that in many instances threshold values had to be	Such consideration does not warrant the assumption that there is a possibility of harm also in the present case

Table 6.1 (continued)

Arguments of Plaintiff	Responses of the Court
lowered because new scientific knowledge revealed that there was in fact a danger	
In view of the uncertainties regarding possible risks the use of nano particles should not be allowed until parliament has determined that it is acceptable	Lack of knowledge about risks does not imply that the use of nano particles is unlawful until it has been explicitly authorized

The fact that it may be impossible to exclude risks, or that unknown risks may exist does not warrant a ban on the technology.

4. Risk evaluation implies risk comparison. Judgements pertaining to the acceptability of a technology must be consistent and non-discriminatory. Comparable technologies must be treated equally.

The judgment of the Administrative Court applies the standard of '*Gefahrenabwehr*' (precluding anticipated adverse effects) rather than '*Risikovorsorge*' (precaution against risks).[3]

However, this logic also applies under the PP. As a legal standard the PP cannot be used to reverse the sequence of risk assessment and risk evaluation or evade the need for empirical risk analysis altogether. The requirement that it must first be established that hazards exist before a technology can be banned because it is too risky is a rule of common sense and is presupposed in all regimes of risk regulation.[4] All these regimes reproduce the classical distinction between statements of fact and value judgements. Evidently, normative criteria precede the empirical analysis by circumscribing which facts count as risks – normally adverse effects in a limited number of protected spheres such as human life and health, or environmental quality. However, whether these effects are likely to occur is an empirical question independent of normative criteria.

The PP lowers the burden of proof for adverse effects, but it does not make it obsolete. The German Federal Administrative Court has ruled in a case concerning nuclear energy that a 'suspicion of danger or merely the potential for concern' suffices to warrant restrictive action under the PP.[5] However, it is not enough merely to entertain a suspicion or concern. Suspicion or concern must be grounded in sufficient reason.[6] The Court acknowledged that 'merely theoretical considerations and calculations' can constitute sufficient reason, but it did not accept 'that a minority opinion in science advances a risk hypothesis that the majority cannot refute on the basis of existing experimental and theoretical knowledge'.[7] Neither is it enough that there could be hidden risks that cannot be detected with available knowledge, and which therefore can neither be predicted nor excluded with a degree of certainty. The Federal Constitutional Court (Bundesfassungsgericht) has relegated such uncertainty to the category of 'residual risk', which can never be excluded and is acceptable in society. Any other approach would mean that reference to the PP would render nuclear energy automatically inadmissible under all circumstances.[8]

The PP does not mandate regulatory agencies to reverse the burden of proof completely. Such a reversal would imply that risk assumptions no longer have to be substantiated by critics of technology (or the regulatory agency), but be refuted by its proponents. Any suspicion or concern would have to be taken seriously until proven unfounded. The philosopher Hans Jonas argued that the reversal of the burden of proof followed from the principle of responsibility (Jonas 1979). Where there exists uncertainty about the consequences of our actions we must apply a 'heuristics of fear', which assumes that the worst possible outcome is likely to happen. However, the PP is not a legal manifestation of the heuristics of fear. It only condones a limited reversal of the burden of proof. Most regulations require that the relative safety of a new technology be demonstrated by passing it through a filter of preventive testing before it can be unleashed on society. To that extent, the burden of proof rests indeed with proponents of technology, and if the technology fails to pass the test, it will not be authorized. However, proponents must only provide the evidence that is specified in the regulation. They must, so to speak, refute hypothetical risks that have already been substantiated and operationalized. They must neither prove that the regulatory requirements are adequate, nor that the regulation does, in fact, consider all possible risks. Neither do they need to show that no risks exist, which we do not or cannot yet know. Such a burden would automatically exclude the technology under all circumstances, which is beyond the legal rationale of authorizing the technology within a PP framework.

The PP does not exempt competent authorities from the need to apply a comparative approach in risk evaluation. While the principle provides a mandate to take uncertainties about risks into account, and hence impose additional restrictions and safety measures, these restrictions and measures must be consistent with existing risk regulation. Thus, it would hardly be possible to use the PP to impose zero emission standards on a nano technology production site in order to forestall any possible adverse effects on the neighbour's health, when at the same time emissions of nano particles from diesel-fuelled automobiles are considered acceptable under the PP.

Interpretations of the PP that cannot be defended as rules cannot be sustained in the law.

As a legal principle, the PP cannot be used to discriminate against a specific technology. Precautionary standards developed for one case must be applicable to comparable cases and must yield reasonable results for such cases. The imperative of consistency also operates against an interpretation of the PP that requires zero risk or reverses the burden of proof completely, because if such requirements were applied no new technology could ever pass the hurdle of the PP.

The conclusion is that regulation under the PP is less an alternative to the liberal regime of innovation than a variant of it. The PP allows more restrictive regulation of new technology by taking remote possibilities of harm into account. However, it does not provide a carte blanche to reject a technology that gives rise to unsubstantiated fears, or because it is undesirable for other reasons. The PP is not designed to subject the dynamics of innovation to democratic political control, and if there are insufficient reasons to suspect risks, new technology will be released into the society. Thus, even with the PP in place, the regulation of new technology still proceeds from the premise that the possibility of risks that we do not or cannot know are a price worth paying for innovation, and that we will somehow be able to cope with unforeseen consequences of the technology if and when they emerge.

This interpretation is, however, only valid as long as the PP is applied as a legal principle and as long as its application has not become a battlefield for social conflicts over new technologies. These conditions may (still) hold in the case of nano technology and perhaps mobile telecommunication (but see Burgess 2003). They do not hold, apparently, in the case of genetically modified crops (GMCs) in Europe. In this latter case, the PP has become politicized and been turned into a means to suppress innovation not because it was too risky, but because it was otherwise deemed undesirable.

3 RISK PREVENTION OR POLITICAL CONTROL OF INNOVATION

The possible risks of GMCs have been the subject of numerous experiments, analyses and reports during the past two decades, all of which came to the conclusion that risks of GMCs are indistinguishable from those related to new crops that are conventionally bred. This 'normalization through comparison' not only applies to recognizable risks, such as cross-pollination from new varieties or accumulation of toxic substances within the plant, it also holds true for uncertainties, lack of data, and the limits of our knowledge, to which critics routinely refer to justify a ban on GMCs. Conventional breeders, too, have never been able to predict with certainty the physiological or phenotypical effects new genes might have, given the genetic background of the plant with which they are crossed. Unexpected and undesirable side-effects abound in breeding and must always be coped with by testing the outcome, and selecting those examples that are suitable for crop development. Lack of knowledge and the limits of testing methods and testing capacities confront conventional breeders in the same way as they do plant geneticists. With respect to possible long-term impacts of new crops on the evolution of species and ecosystems, one can only say that these are as indeterminate and unforeseeable for conventionally-bred crops as they are for transgenic crops.

Those who argue that, in view of such risks and uncertainties, GMCs must be banned under the PP, should, if they are to remain consistent, similarly argue that conventionally bred crops must be banned. Of course, nobody would consider such a ban on conventional crops. However, GMCs were effectively banned within the European Union (EU) in the late 1990s. Any inconsistency was avoided by assuming that, despite lack of evidence, GMCs are in fact different from conventional crops and constitute, therefore, a new type of potential hazard. This position resonated well with the opinion of many people who rejected GMCs and were concerned that they might pose a new kind of risk. In addition, governments were under pressure from organizations such as Greenpeace who crystallized and amplified public protest through massive campaigning. In the end, the public concern became sufficient justification to impose a ban on GMCs, and whether there was sufficient ground for concern was no longer a question. This result was achieved by removing science from the domain of risk policy, which became particularly apparent in Germany where the competent authorities were responsible to the Health Ministry dominated by the Green Party. The case of GMCs illustrates how the PP can be used to restrict innovation on the mere basis of risk perception – without risk assessment to back up such decisions.

However, the ban on GMCs ran into difficulties, since there existed national and EU law that envisaged authorization of GMCs after risk assessment, and which does not allow a precautionary ban without justification. Governments dealt with this problem by withholding the authorization for commercial uses of GMCs through an extra-legal moratorium. When these Member States could not prevent the European Commission to authorize GMCs, they explicitly refused to implement these decisions at national level. Legally, it would have been possible to take these governments to the European Court of Justice (ECJ) for breaches of European Community (EC) law. Politically, however, such a move would have been extremely unpopular, and neither the European Commission nor any other affected party has pursued that route.

The politics of precaution in the EU is clearly illustrated by the story of Novartis' Bt-176 maize. This maize was genetically modified to express an insecticidal gene that makes the plant resistant against a major maize pest. It also contained a marker gene conferring resistance against the antibiotic ampicillin. Novartis applied for a marketing authorization in France in 1995. The responsible French authority forwarded the application with a favourable opinion to the European Commission for European authorization. What follows is a short chronology of what happened afterward (see also Hervey 2001).

April 1996: the European Commission proposes to authorize Bt-176, but several Member States (Austria, Denmark, Germany, Sweden and the United Kingdom) raise objections. They refer to, inter alia, new information that calls into question the safety of the marker gene.

December 1996: since the Member States fail to settle the issue in either the regulatory committee pursuant to Article 21 of Directive 90/220/EEC on the Deliberate Release into the Environment of Genetically Modified Organisms (Deliberate Release Directive 1990), or in the European Council, the case is referred back to the Commission, which seeks advice from its three scientific committees (food, animal nutrition, pesticides) on the new information invoked regarding the safety of Bt-176. All three committees conclude that there is no reason to assume that Bt-176 may have adverse effects on humans or the environment.

January/February 1997: the European Commission authorizes the marketing of Bt-176 within the EU. In response, Austria and Luxemburg issue a national ban on Bt-176 based on Article 16 of the Deliberate Release Directive, which allows unilateral national measures when new information suggests that GMCs are unsafe. Austria also points out that one-fifth of its population has signed a petition against Bt-176.

September 1997: the Commission introduces the infringement procedure of Article 216 EC to ensure compliance with the Deliberate Release Directive, and orders Austria and Luxemburg to lift their bans on Bt-176. Before doing so, the Commission had again asked its scientific committees to review the new information, and received confirmation that no arguments had surfaced indicating that Bt-197 might pose risks to human health. However, the procedure initiated by the Commission does not give rise to any further action, either in the responsible EU committee, or in the European Council. Instead, Member States discuss the need to revise the Deliberate Release Directive.

June 1999: the European Council of Environmental Ministers agree on a common position to amend the Deliberate Release Directive so as to emphasize the PP. However, they do not impose a formal moratorium, because that would have been in breach of EC and national law. Instead, they informally agree that no further GMCs should be authorized at the EU level.

February 2000: the German Federal Office of Plant Varieties (Bundessortenamt) announces that it will accept the registration of Bt-176 as a variety, thus satisfying the final legal requirement allowing the marketing of Bt-176 in Germany. The Health Ministry immediately orders the Robert Koch Institute (RKI), responsible for authorizing GMCs, to revoke the European market authorization for Bt-176 on the ground of Article 16 of the Deliberate Release Directive, that is that new information suggests that it is unsafe. The Ministry based this on new studies, among which was a report it had commissioned to the Öko-Institut, a group of academics closely affiliated with the environmental movement pursuing, amongst others, the critical observation and monitoring of genetics.

March 2000: in a meeting with the Health Ministry, the Central Commission for Biological Safety (ZKBS), which is the scientific advisory body for the RKI, distances itself from the procedure chosen by the Ministry. It points out that the Ministry had already received reports from the ZKBS in which the information presented as new had been evaluated and no reason was found to suspect adverse effects from Bt-176. The ZKBS criticizes the Health Ministry, which it claims had abandoned sound science by disregarding the advice of the ZKBS, and relying on the opinion emanating from the Öko-Institut instead. The Ministry counters that the revocation of the authorization is a value judgement, not merely a scientific assessment.

September 2000: the European Commission scientific committee on pesticides reports on the allegedly new information (including that from the Öko-Institut) invoked by the German government to justify the national withdrawal of the Bt-176 authorization. It concludes that there is no reason to deviate from previous evaluations that confirm the acceptability of Bt-176.

The use of antibiotic resistance genes as markers in GMCs, although considered safe, was hardly ever accepted as appropriate technology. Article 4(2) of Directive 2001/18/EC on the Deliberate Release into the Environment of Genetically Modified Organisms and Repealing Council Directive 90/220/EEC (Deliberate Release Directive 2001) now limits such use until 2008. Moreover, in April 2004 the European Food Safety Agency issued a recommendation to ban the use of the ampicillin resistance marker genes in commercially cultivated GM crops (GMO Panel 2004).

The case history of Bt-176 is illustrative, as it shows that various governments in the EU not only abandon established science as a basis for risk assessment, but also breach prevailing law to pursue a ban on GMCs. They feel compelled and legitimized by critical public opinion and occasionally parliamentary initiatives. Reportedly, the German Health Ministry has bluntly argued that in such situations 'political will must have the precedence over existing law' when it met with resistance from national competent authorities who insisted that the law does not warrant revocation of the Bt-176 authorization.

While such a statement is astonishing coming from an executive body that the German constitution explicitly submits to law and justice,[9] it is beyond doubt that in making the law, legislatures, both at the European and Member State level, enjoy a high degree of discretion to impose restrictions on new technologies that they deem necessary for reasons of risk precaution. The Federal Constitutional Court of Germany recently confirmed that parliament – after balancing private and public concerns – can impose precautionary measures, even if these appear to come 'out of the blue' ('*ins Blaue hinein*').[10]

If, however, a regulation renounces all requirements to substantiate reasons for risk assumptions, and precludes the control of such assumptions through scientific assessment of cause–effect relationships, it would be more honest to speak of political planning rather than of risk precaution. In such cases, a ban on GMCs should be legitimized on the ground that many people fear risks, and therefore prefer to have genetic engineering kept out of agriculture and food production (van den Daele 1999). In fact, political planning seems to be the hidden agenda in many GMC regulations in EU countries. One example is Austria, which tenaciously pursues the goal of GM-free agriculture, even at the cost of breaching EC law. Elements of political planning are also implied in the rules that complement GMC regulations to ensure the coexistence of conventional, organic and GM-based farming and food production.[11] Policies of risk prevention clearly merge with political planning of technology if the criteria of the harm to be avoided are extended beyond adverse effects on health and environment to include social, cultural and political effects. Thus, according to Section 63 of the Austrian Gene Technology Act of 1994, the government must prohibit products of genetic engineering that 'could lead to a burden on society or social groups, which cannot be compensated, and if the burden on the society appears unacceptable for economic, social or ethical reasons'.[12] Similarly, Section 10 of the Norwegian Gene Technology Act of 1993 requires that 'in deciding whether or not to grant the application, significant emphasis shall be placed on whether the deliberate release represents a benefit to the community and a contribution to sustainable development'. The Danish government has promised an effort to include ethical criteria into European and international regulation that would make the authorization of GMCs contingent upon the equal distribution of burdens and benefits in society.[13]

4 THE RETURN OF PRECAUTION AS A LEGAL PRINCIPLE

We will skip the question as to the extent to which introduction of new technology may theoretically be placed under political planning in liberal societies. Despite the rhetoric just cited, it seems unlikely that modern states will really engage in comprehensive political planning to determine which technology will be used in society, applying, for example, the criteria of ecological, social and economic sustainability. Political preferences may favour organic farming in some countries, but that has only led to promotion of this type of agriculture, basically adding to other technological options available in the society. No government has attempted to cut back conventional agriculture merely on the grounds that it does not comply with the criteria of organic farming. Restrictions on new technology continue to be justified in terms of precautionary risk prevention, not in terms of political planning. This will also apply to GMCs.

Regulation in terms of risk prevention fits the present political landscape much better than regulation in terms of social planning. In view of the widespread scepticism regarding state control of society, it is easier to justify a ban on new technology with the argument that the ban is necessary to protect health and the environment, than by simply saying that the ban has been decided by a majority of society. Moreover, risk prevention becomes an almost unavoidable frame for the regulation of technology because states have reduced leeway for political planning through international trade regimes. Under the laws of the World Trade Organization (WTO) and even more so under the EC Treaty, a national law that bans GMCs simply to keep the country GM-free would amount to protectionism and would have as little chance of being upheld as a law banning US computers or Japanese cars. Only if the regulation can somehow be brought under the umbrella of protection can it escape the verdict of protectionism.[14]

However, within the EU it has become difficult to advance an ambiguous PP to justify regulation that, in fact, pursues political ends other than risk prevention; for example, to impose a ban on GMCs because the technology is rejected by a sizeable part of society, or because it is does not conform to a preferred image of how agriculture should develop. The Communication of the European Commission on the Precautionary Principle (COM 2000), the Deliberate Release Directive 2001, and most notably the ECJ, militate against a PP that operates as a political bargaining chip to be used for a multitude of purposes, and have re-established it as a legal principle that can only be invoked if definite prerequisites are fulfilled. The following quotations are from the Judgment of the Court of First Instance (CFI).[15]

> The European Court has ruled that the PP allows restrictive measures on the ground that there is uncertainty about risks, but only 'if the risks, although the reality and extent thereof has not been fully demonstrated by conclusive scientific evidence, appears nevertheless adequately backed up by the scientific data available' (no. 157). Preventive measures cannot properly be based on a 'purely hypothetical approach to the risk, founded on mere conjecture' (no. 156). The PP does not justify to call for zero risk 'since it is not possible to prove scientifically that there is no current and future risk' (no. 158). Regulatory agencies enjoy a broad discretion in determining what counts as ground for concern; they may also rely on scientific opinion which is contested in the respective scientific community. But they have in any case to seek scientific advice that satisfies the principles of 'excellence, independence and transparency' (no. 183). This is an indispensable procedural requirement to ensure that preventive action is based on the best available scientific data and the most recent results of international research and does not adopt arbitrary measures 'which cannot in any case be rendered legitimate by the precautionary principle' (no. 175). Under these rules it would clearly be impossible for the German government to brush all established science bodies aside and base its finding that there is sufficient ground for concern on the opinion from a think tank of the environmental movement. (Öko-Institut)

The Communication on the Precautionary Principle points out that, as a legal rule, the PP must be applied consistently and cannot be used to discriminate against a specific technology. While legislators may enjoy discretion to impose inconsistent regimes on different technologies (insofar as this is compatible with international trade agreements), regulatory agencies cannot justify a ban of GMCs based on risks and uncertainties, which also exist for conventional crops, but which are considered acceptable in that field. Likewise, it would be inappropriate to apply criteria of sustainability or standards of organic farming as a frame of reference in the evaluation of the environmental effects of GMCs, at least as long as the effects of conventional agriculture are not evaluated accordingly (Levidow et al. 2000, p. 196).

In sum, the interpretation of the PP incorporated in EC law has become de-politicized.[16] The PP cannot be used to translate fears or preferences in public opinion into restrictive measures without further justification (Hervey 2001). Only legislators have the political power to declare without further proof that the possible risks of antibiotic resistance genes are unacceptable, and that therefore these genes cannot be authorized as markers in GMCs. They may also ban such markers to alleviate concerns of the public or forestall protest, particularly when alternative marker techniques are available. Article 4(2) of the New Deliberate Release Directive, which limits the use of such marker genes until 2008, can be regarded as a product of the exercise of such political power. But regulatory agencies, in imposing preventive measures under the PP, have no similar mandate. They must substantiate the possibility of harm, and cannot have recourse to political judgement when evidence of risk is lacking. Nor can they, in the name of the PP, forego consultation with established science and choose 'democratization' of expertise instead.[17] Finally, the PP does not allow discrimination against GMCs in evaluating the acceptability of risks by applying standards of social justice or sustainability which are not applied when conventional agricultural practices are regulated.[18]

5 ONGOING BATTLES

Official EU rhetoric is that with the New Deliberate Release Directive and related EU legislation (on labelling, monitoring, traceability and so on) in place, the moratorium on GMCs that has existed since 1998 has now been lifted. This does not necessarily mean that the blockade of GMCs within the EU has effectively ended. The German Federal Health Ministry decided in 2000 to boycott GM agriculture in Germany even with the New Deliberate Release Directive in place. An internal policy assessment was confident that the consolidated new rules provide space to delay authorizations considerably through filibustering. It emphasized that public acceptance of GMCs was still low, and the resistance of social movements undiminished. The paper saw no prospect that Greenpeace and other environmental groups would compromise after they had achieved what they wanted, that is, that GM agriculture has no future within the EU. The Ministry also felt that the majority of the German Federal Parliament (Bundestag), was at least tacitly in favour of continuation of the blockade of GMCs. Within the government coalition, only some Social Democrats and the Federal Ministry of Economics were still considered to be advocates of GMCs.

One can expect that EU Member States that strive to keep GM agriculture out of their territory will continue to block authorization of GMCs, arguing that in view of the lack of data and of yet undiscovered long-term risks, such crops should be banned under the PP. This policy is fuelled by social movements that crystallize public mistrust in existing regulation through direct actions, including environmentalists wearing protective clothing designed for biological warfare while 'harvesting' experimental GM crops. These actions are extensively covered in the mass media, and feed into risk perceptions that associate GMCs with dangerous toxins. In November 2004, the European Commission failed to reach a qualified majority in the Regulatory Committee on the Deliberate Release of GMOs for its proposal to force five Member States (Austria, France, Germany, Germany, Luxembourg) to lift national bans that they had imposed on GMCs or products thereof despite authorization on the EU level. The Regulatory Committee represents the Member States. Apparently, quite a few governments shared the feeling of the non-governmental organization (NGO) Friends of the Earth that the Commission's proposal was 'deeply unpopular and clearly undemocratic' (Friends of the Earth 2004). National laws on nature conservation and the coexistence between GM agriculture and organic farming seem particularly apt places to impose a ban on GMCs, despite authorization in Brussels. In October 2004, the Parliament of Kärnten in Austria unanimously passed a law that requires a 3-kilometre distance between GM crops and nature reserves or organic farming sites, which means that the territory is kept GM-free as a result. It remains to be seen whether seed companies and the food industry will apply for authorization to market GMCs under these conditions. It is not unrealistic to expect that they would rather retreat from GM technology for agriculture within the EU. The implementation of the New Deliberate Release Directive into German law has added further hurdles, in particular by imposing strict liability with a reversal of the burden of proof for any economic loss neighbouring farmers may suffer through cross-pollination from transgenic plants.[19]

6 SYMBOLIC USE OF POLITICS

Let us assume that the ongoing efforts to uphold the blockade of GMCs and their products in Europe will prove to be successful. What kind of political result would that be? It would certainly show that the self-propelled dynamics of technological innovation can be broken (at least regionally). The critics of GM would have defeated the protagonists, and have demonstrated that the liberal regime of innovation can be reversed if resistance is mobilized in society, and that governments and industries can be forced to move beyond the established safety controls and renounce a technological opportunity because it is not sufficiently accepted. One may consider the example an important achievement per se. But what are the political benefits in terms of the substantive goals the blockade is supposed to serve? Have we achieved better health and a better, safer environment? Are we any step nearer to reinstating the primacy of democratic rule over the dynamics of technological change? In both respects, the effects are questionable and in any case trivial compared with the real problems we are facing. The blockade of GMCs amounts rather to the symbolic use of politics.

The ban on GMCs is insignificant as a measure of precaution. In fact, it only reduces the residual risk that there could be adverse effects on human health and the environment, which so far can neither be backed by empirical evidence, nor theoretically anticipated, and which therefore escapes established precautionary controls of new technology. The ban ends the invasion of GMCs into conventional agriculture. This averts potential risks of which nobody can say they exist. Conversely, conventional agriculture is fraught with manifest, recognizable risks such as soil erosion and salination, euthrophication of water, and loss of biodiversity through high yield crops. None of these problems is in any way addressed through the ban on GMCs. Quite the contrary, the ban precludes possible options to improve the environmental record of conventional agriculture. Against this background, it is probably fair to say that the ten-year battle that has been fought over GMCs between political parties, governments, agencies and numerous actors of civil society was a waste of scarce political resources.

It would also be misleading to take the ban on GMCs as proof that politics has asserted its sovereignty over the dynamics of technology. First, the blockade of GMCs is confined to a couple of EU Member States and holds only for the cultivation of GM crops but not for the import of products from GMCs, as is illustrated by the wide use of GM feed in all European countries. Second, and more important, GMCs represent a marginal factor in the wave of technological innovation sweeping through modern societies. A revolution is under way in the information and communication technologies with dramatic effects on the living and working conditions of millions of people. Here it becomes apparent that politics can only follow the technological dynamics that are unleashed in liberal-capitalist societies, and has neither the will, nor the power to subject it to democratic control.

Finally it should be clear that the blockade of GMCs is a precarious policy that may be revoked any time. It is mainly legitimized by the fact that the majority of people reject the products of GMCs. However, this rejection has been drawn from opinion polls. Whether people will actually reject GM products when they can choose to buy them in the supermarket is a different matter. Resistance may well break down once products become available that provide real benefits, not only to farmers and food industries, but also the environment and consumers. For many years genetically modified micro-organisms (GMOs), too, have been fought with the argument that they may bear undetected, long-term risks and that the regulation cannot be trusted. None of these arguments continued to carry weight after the first medicines from GM organisms had reached the patients. For GMCs a similar shift may be anticipated, if useful products are developed. If the blockade of GMCs continues, such products will have to be imported into Europe because they could not be developed here.

NOTES

1. The precautionary principle (PP) is articulated in Principle 15 of the Rio Declaration (Rio Declaration on Environment and Development 1992) as follows: 'Where there are threats of serious or irreversible damage, lack of full scientific certainty shall not be used as a reason for postponing cost-effective measures to prevent environmental degradation'. For another version of the PP, found in Article 10.6 of the Cartagena Protocol (Cartagena Protocol on Biosafety to the Convention on Biological Diversity 2000), it has been suggested that it is a binding rule of international law: 'Lack of scientific certainty due to insufficient relevant scientific information and knowledge regarding the extent of the potential adverse effects of a living modified organism on the conservation and sustainable use of biological diversity in the Party of import, taking also into account the risks to human health, shall not prevent that Party from taking a decision, as appropriate, with regard to the import of the living modified organism . . . in order to avoid or minimize such potential adverse effects'. For a discussion of the rise of the PP see Freestone and Hey (1996); Harremoës et al. (2002); Raffensberger and Tickner (1999).
2. Arguments in square brackets in Table 6.1 are inferred from the responses of the Court. The decision has been confirmed in Bundesverwaltungsgericht (2003).
3. According to Para. 5, Sect. 1, No. 1 of the Bundesimmissionsschutzgesetz (Federal Emission Control Law). While the precaution against risks standard requires less substantiation of the possibility of harm, the Court doubted whether this would have changed its judgement, because the fears raised by the plaintiffs might belong to the category of 'residual risk', which is irrelevant under both standards (*Die Öffentliche Verwaltung (Public Administration)* (2002) **55**(20), 871 at 874).
4. For instance, the European Council Resolution on the precautionary principle stating that 'the scientific assessment of the risk must proceed logically in an effort to achieve hazard identification, hazard characterisation, appraisal of exposure and risk characterisation' (European Council 2000).
5. The full text reads: 'possibilities for harm must be taken into account that cannot be excluded because, given the present state of knowledge, certain causal relations can neither be confirmed nor denied and hence there is no clear and present danger, but only the suspicion of danger or merely the potential for concern'. See Decisions of the Bundesverwaltungsgericht (1985).
6. In a case concerning atmospheric pollution, the Court required that 'there [be] *sufficient reason* [emphasis added] to assume that emissions could possibly lead to harmful effects on the environment'. See Bundesverwaltungsgericht 1984. In a similar vein the United States Toxic Substances Control Act requires a 'reasonable basis' to assume risk: see United States Code, 15, § 2605 (a). The formulation in the Dutch law is 'a grounded suspicion of undesired effects' ('*een redelijk vermoeden is gerezen van ongewenste effecten*'). See Wet milieugevaarlijke stoffen 5 December 1985. The Verwaltungsgericht Hannover (Administrative Court) 1991 rejected the argument that corn dust was potentially dangerous because more knowledge could reveal that what seems harmless today may in fact prove to be very dangerous. The plaintiff had argued: 'with respect to dioxins, too, the risk potential had not been realized until a few years ago'. The court ruled that the PP, while expected to prevent unknown risks, still necessitates substantiation of real potential for danger (Feldhaus 1998).
7. Decisions of the Bundesverwaltungsgericht (1985), pp. 315 and 318. Otherwise, the Court argued, the consequence would be that under the precautionary principle one had

'to conclude from the ongoing dialogue essential to the scientific process that the peaceful use of nuclear energy is inadmissible'.

8. Decisions of the Bundesverfassungsgericht (1978). If the goal was to exclude the technology under all circumstances the law should have imposed an unconditional ban, but in fact the Atomgesetz (Atomic law) authorizes the use of the technology provided adequate precautionary measures are taken.
9. According to Art. 20, Sect. 3 of the German *Grundgesetz* (Basic law) the executive shall be 'bound by law and justice' (*an Gesetz and Recht gebunden*).
10. Bundesverfassungsgericht (2002). 'Out of the blue' ('ins Blaue hinein') refers to an influential article by Ossenbühl, who warned that the inclusion of the PP in legal regulations must not be understood as empowerment of the administration to impose restrictions without justification 'ins Blaue hinein' (Ossenbühl 1987). This warning is valid for the application of law, but it does not apply to law-making. In a judgement of 3 July 2002, the German Federal Administrative Court, too, underlined that legislators can restrict private rights (in this case the right to keep dogs) 'to minimize risks for which regulation is demanded either because of new reasons to suspect risks or because of changes in the society or of shifting perceptions of the people'. Cf. Decisions of the Bundesverwaltungsgericht (2002).
11. Guidelines for such rules that can be promulgated by the Member States have been proposed by the European Commission in Commission Recommendation of 23 July 2003.
12. It is questionable whether this rule is admissible under the Austrian Constitution (Waldhäusl 1995).
13. 'Genetic engineering must be used in such a manner that it does not conflict with our efforts to create a society where benefits and burdens are distributed equitably. This consideration applies both within the individual society as well as with regard to fostering sustainable development in relation to other countries, including developing countries, and in relation to future generations' (BioTIK Action Plan 2001).
14. Thus, Art. 2(2) of the Agreement on Sanitary and Phytosanitary Measures of 1994 (SPS Agreement 1994) reads: 'Members shall ensure that any sanitary or phytosanitary measure is applied only to the extent necessary to protect human, animal or plant life or health, is based on scientific principles and is not maintained without sufficient evidence . . .' Temporary measures may be taken according to Article 5(7) 'in cases where relevant scientific evidence is insufficient'. Even if under WTO law regional planning may be a legitimate means to exclude the cultivation of GMCs in a particular country, it would not allow restrictions on imported GMCs from other countries.
15. Judgement of the Court of First Instance (CFI) in Case T-70/99, 2002.
16. With respect to the de-politicization pushed for by the Commission Communication of 2000 one observes submission to the international trade regimes of GATT/WTO (Appel 2003). This can hardly be maintained with respect to the interpretation of the PP by the ECJ.
17. However, according to Theofanis Christoforou, a staff member at the European Commission, the agencies, too, enjoy political discretion that may complement or replace scientific assessment whenever it is necessary to determine what available scientific data mean for regulation, or whether there is the possibility of harm and what kind of harm this could be. He proposes that 'risk managers', instead of trying to patronize consumers with positivist views on science, 'should take into account their [the consumers's] legitimate concerns and the public's perception of risk' (Christoforou 2003, p. 208).
18. The Deliberate Release Directive (2001) reserves no role for economic and social effects in the procedures for authorizing GMCs. The Commission is merely advised to monitor such effects and furnish a report on them (Recital 62).
19. Gentechnikgesetz (GenTG) (1990) and (1993). Cross-pollination is unavoidable and common in agriculture. Events that are bound to happen with certainty are not 'risks' within the framework of insurances, therefore no coverage will be offered for such liability. In Germany, critics have argued that the moratorium on GMCs should not be lifted unless liability laws are in place that shift the burden of proof to GM farmers and a collective fund is created to cover all damages that might occur (Hermann and Tappeser 2004).

REFERENCES

Agreement on the Application of Sanitary and Phytosanitary Measures (SPS Agreement) (1994), Marrakesh Agreement Establishing the World Trade Organization, Annex 1A, Legal Instruments – Results of the Uruguay Round Vol. 1 (1994), *International Legal Materials*, **33**, 1125–53.

Appel, I. (2003), 'Präventionsstrategie im europäischen Chemikalienrecht und Welthandelsrecht', *Zeitschrift für Umweltrecht*, **14** (Special Issue No. 1), 167–75.

Beck, Ulrich (1992), *Risk Society: Towards a New Modernity*, London: Sage.

BioTIK Action Plan (2001), 'Ethics – A Tool for Making the Right Choices on Biotechnology', http://www.biotik.dk/english/plan/cer/.

Bundesimmissionsschutzgesetz (Federal Emission Control Law) (BImSchG) *Bundesgesetzblatt* (Federal Law Gazette) (BGBl), **I** (1990), 880.

Bundesverfassungsgericht (Federal Constitutional Court) (1978), Bundesverfassungsgericht 8 August 1978 (Kalkar), Entscheidungen (Decisions) 49, 89, at 143.

Bundesverfassungsgericht (Federal Constitutional Court) (2002), Bundesverfassungsgericht 28 February 2002, *Die Öffentliche Verwaltung* (*Public Administration*), **55**(12), 521.

Bundesverwaltungsgericht (Federal Administrative Court) (1984), Bundesverwaltungsgericht 17 February 1984 (Heizkraftwerk) (Combined heat and powerstation), Entscheidungen (Decisions) 69, 37 at 43.

Bundesverwaltungsgericht (Federal Administrative Court) (1985), Bundesverwaltungsgericht 19 December 1985 (Whyl), Entscheidungen (Decisions) 72, 300 at 315.

Bundesverwaltungsgericht (Federal Administrative Court) (2002), Bundesverwaltungsgericht 3 July 2002, Entscheidungen (Decisions) 116, 347, at 353.

Bundesverwaltungsgericht (Federal Administrative Court) (2003), Bundesverwaltungsgericht 11 December 2003, (2004) *Die Öffentliche Verwaltung* (*Public Administration*), **57**(8), 340–43.

Burgess, Adam (2003), *Cellular Phones, Public Fears, and a Culture of Precaution*, Cambridge, UK: Cambridge.

Cartagena Protocol on Biosafety to the Convention on Biological Diversity (2000), *International Legal Materials*, **39**(5), 1027.

COM (2000), 'Communication from the Commission on the Precautionary Principle', *Commission of the European Communities*, **1**, Brussels, 02-02-2000.

Christoforou, T. (2003), 'The Precautionary Principle and Democratizing Expertise: A European Legal Perspective', *Science and Public Policy*, **30**(3), 205–13.

Commission Recommendation of 23 July 2003, *Official Journal of the European Union*, L 189/36, 2003/556/EC.

Deliberate Release Directive (1990), 'Directive 90/220/EEC on the Deliberate Release into the Environment of Genetically Modified Organisms', *Official Journal of the European Communities*, L117/15.

Deliberate Release Directive (2001), 'Directive 2001/18/EC on the Deliberate Release into the Environment of Genetically Modified Organisms and Repealing Council Directive 90/220/EEC', *Official Journal of the European Communities*, L106/1.

European Council (2000), 'Council Resolution on the Precautionary Principle', *Bulletin EU 12-2000 Annexes to the European Council Conclusions 4/7*, Annex III.

Feldhaus, Gerhard (1998), *Bundesimmissionsschutzrecht: Entscheidungssammlung* (Federal Pollution Protection Law: Law Reports), Volume 3, § 5, No. 38, Heidelberg: C.F. Müller.

Freestone, David and E. Hey (1996), *The Precautionary Principle and International Law: The Challenge of Implementation*, The Hague: Kluwer.
Friends of the Earth Press Release (29 November 2004), 'Europe Votes to Keep GM Crop Bans', http://www.gmwatch.org/archive2.asp?arcid=4674.
Gentechnikgesetz (GenTG) of 20 June 1990 BGBl I 1990, 1080 as amended 16 December 1993 – BGBl, I S. 2066.
GMO Panel (2004), 'Opinion of the Scientific Panel on Genetically Modified Organisms on the Use of Antibiotic Resistance Genes as Marker Genes in Genetically Modified Plants', *European Food Safety Authority Journal*, **48**, 1–18.
'Grundgesetz für die Bundesrepublic Deutschland' (Basic Law for the Federal Republic of Germany, Bundgesetzblatt (Federal Law Gazette)) III 100–101, 23 May 1949.
Harremoës, Poul, D. Gee, M. McGarvin, A. Stirling, J. Keys, B. Wynnie and S. Guedes Vaz (eds) (2002), *The Precautionary Principle in the 20th Century. Late Lessons from Early Warnings*, London: Earthscan.
Hermann, A. and B. Tappeser (2004), 'Entwurf des Dritten Gesetzes zur Änderung des GenTG vom 16.01.2004. Bewertung der Änderungen im GenTG unter besonderer Berücksichtigung der Koexistenz in der Landwirtschaft. Gutachten des Ökoinstituts vom Februar 2004', http://www.oeko.de.
Hervey, T.K. (2001), 'Regulation of Genetically Modified Products in a Multi-level System of Governance: Science or Citizens?', *Review of European Community and International Environmental Law*, **10**(3), 321–33.
Jonas, Hans (1979), *Das Prinzip Verantwortung. Versuch einer Ethik für die technologische Zivilisation*, Frankfurt am Main: Suhrkamp.
Judgement of the Court of First Instance (CFI), 11 September 2002/ C 289/41, 'Case T-70/99 *Alpharma Inc.* v. *Council*', *Official Journal of the European Communities*, 23 November 2002, ECR II-3495.
Levidow, L., S. Carr and D. Wield (2000) 'Genetically Modified Crops in the European Union: Regulatory Conflicts as Precautionary Opportunities', *Journal of Risk Research*, **3**(3), 261–70.
O'Riordan, Timothy and James Cameron (eds) (1994), *Interpreting the Precautionary Principle*, London: Earthscan.
Ossenbühl, F. (1986), 'Vorsorge als Rechtsprinzip im Gesundheits-, Arbeits- und Umweltschutz', *Neue Zeitschrift für Verwaltungsrecht*, **5**(3), 161–71.
Raffensberger, Carolyn and Joël Tickner (eds) (1999), *Protecting Public Health and the Environment: Implementing the Precautionary Principle*, Washington, DC: Island Press.
Rio Declaration on Environment and Development (1992), *International Legal Materials*, **31**, 874.
Slovic, Paul (2000), *The Perception of Risk*, London: Earthscan.
Tait, J. and L. Levidow (1992), 'Proactive and Reactive Approaches to Regulation: The Case of Biotechnology', *Futures* (April), 219–31.
United States Code, 15, § 2605 (a).
van den Daele, Wolfgang (1999), 'Von rechtlicher Risikovorsorge zu politischer Planung: Begründungen für Innovationskontrollen in einer partizipativen Technikfolgenabschätzung zu gentechnisch erzeugten herbizidresistenten Pflanzen', in Alfons Bora (ed.), *Rechtliches Risikomanagement: Form, Funktion und Leistungsfähigkeit des Rechts in der Risikogesellschaft*, Berlin: Dunckler & Humblot, pp. 259–91.

Verwaltungsgerichtshof of Baden-Württemberg (2002), 'Verwaltungsgerichtshof Baden-Württemberg 18 December 2002', (2002) *Die Öffentliche Verwaltung*, **55**(20), 871–5.

Waldhäusl, Martin (1995), 'Soziale Unverträglichkeit – Rechtlicher Rahmen und Anknüpfungspunkte in der Österreichischen Rechtsordnung', in Helge Torgersen and Franz Seifert (eds), *Die Sozialverträglichkeitsprüfung von gentechnischen Produkten zwischen Anspruch und Umsetzbarkeit*, Vienna: Institut für Technikfolgen-Abschätzung der Österreichischen Akademie der Wissenschaften.

Wet milieugevaarlijke stoffen, 5 December 1985, Art. 24, Stb. 639, last amended 18 October 2001, Stb. 517.

Wynne, B. (1992), 'Uncertainty and Environmental Learning: Reconceiving Science and Policy in the Preventive Paradigm', *Global Environmental Change*, **2**(2), 111–27.

7. Regulating GM food: three levels, three issues

Bernd van der Meulen*

1 INTRODUCTION

This chapter focusses on the regulatory framework for genetically modified organisms (GMOs), including micro-organisms, used for human consumption (GM foods). Environmental considerations associated with GM crops will not be considered here.

In Europe, European Community (EC) law provides the core of the regulatory framework on GM food. EC law does not, however, tell the whole story. National law on the one hand and international law (WTO, Codex Alimentarius) on the other are also of key importance.

The subtitle of this chapter refers to the global, the European and the national levels. However, our focus will be on the European level, with particular emphasis on three procedural issues: approval, handling and labelling. Issues of national and international law will only be touched upon in passing.

'Handling' and 'labelling' concern all food business operators in the EU. We shall discuss the rules such operators have to comply with when they are – or may be – confronted with GM foods. The issue of approval is relevant for the innovative industry bringing new GM food products onto the market. As regards approval, the predominant question is: 'Who is in charge?' The most sensitive issues, however, concern the consequences of this legal regime for primary production. Finally, international law raises the question of whether the EU regulatory framework is likely to have to undergo major changes in the near future.

2 EUROPEAN LAW ON GM FOODS

2.1 Overview

Triggered by the food safety scares of the 1990s, European food law is currently undergoing a transition from being predominantly market-oriented towards a focus on consumer protection, that is, food safety law (Van der Meulen 2004; Van der Meulen and Van der Velde 2004). The European Commission presented its grand design for updating European food safety law in the White Paper on Food Safety (European Commission 1999). The White Paper envisages over 80 legislative initiatives, three of which concern GM food. These three have been amongst the first initiatives to be implemented.

The transition from market orientation to consumer protection has stimulated a tendency towards centralization. As instruments of EC food law, regulations are replacing directives as the preferred form of legislation (Van der Meulen 2004, p. 5),[1] and uniform EC legislation is thus taking the place of harmonized national law. Executive powers are increasingly concentrated in the hands of the European Commission. Apart from power politics, centralization is believed to allow for food safety problems to be addressed more efficiently. For GM food, another important consideration was to bypass Member States who in practice refused to authorize market clearance. This situation is referred to as the 'de facto moratorium'.

The general principles of the new regulatory framework on food are set out in Regulation (EC) No. 178/2002 Laying Down the General Principles and Requirements of Food Law, Establishing the European Food Safety Authority and Laying Down Procedures in Matters of Food Safety (General Food Law). Specific regulations on GM food entered into force on 18 April 2004.[2] These are Regulation (EC) No. 1829/2003 on Genetically Modified Food and Feed (Regulation on GM Food and Feed OJ L 268/1, 2003), and Regulation (EC) No. 1830/2003 Concerning the Traceability and Labelling of Genetically Modified Organisms and the Traceability of Food and Feed Products from Genetically Modified Organisms and Amending Directive 2001/18/EC (Regulation on Traceability and Labelling, OJ L 268/24, 2003). These new regulations take GM foods out of the scope of the Regulation (EC) No. 258/97 Concerning Novel Foods and Novel Food Ingredients (Regulation on Novel Foods, OJ L 43/1, 1997). The most striking difference between the new and old framework is, first, that simplified procedures for novel foods that are 'substantially equivalent' to conventional foods no longer apply to GM foods. Moreover, products of GM origin must now be labelled as such, even if no protein or DNA is present in the final product.[3]

In summary, the most important pieces of EU legislation that together make up the regulatory regime concerning GM food are:

- the General Food Law, which articulates general principles;
- the Regulation on Food and Feed, which concerns the placing on the market of food and feed products containing or consisting of GMOs,

and provides for the labelling of such products for the benefit of the final consumer;

- the Regulation on Traceability and Labelling, which introduces a uniform EU system to trace and label GMOs and to trace food and feed products produced from GMOs;
- Regulation (EC) No. 641/2004 on Detailed Rules for the Implementation of Regulation (EC) No. 1829/2003 (OJ L 102/14, 2004);
- Commission Recommendation on Guidelines for the Development of National Strategies and Best Practices to Ensure Coexistence of Genetically Modified Crops with Conventional and Organic Farming (OJ L 189/3, 2003);
- Directive 2001/18/EC on the Deliberate Release into the Environment of Genetically Modified Organisms and Repealing Council Directive 90/220/EEC (Deliberate Release Directive), which regulates experimental releases and the placing on the market of genetically modified organisms (OJ L 106/1, 2001).

European law on GM food addresses three core issues: pre-market approval, handling and labelling.[4]

2.2 Pre-market Approval

2.2.1 Principles

The General Food Law articulates 'science-based regulation' and the 'precautionary principle' as its guiding general principles. The Regulation on Food and Feed adds to this the procedural principle of 'one door–one key'.

The General Food Law takes as a general principle that in order to achieve the general objective of a high level of protection of human health and life, food law is based on risk analysis (Art. 6.1). 'Risk analysis' is defined as a process consisting of three interconnected components: risk assessment, risk management and risk communication (Art. 3.10). The work connected with these components is further distinguished in terms of scientific research and policy decisions, each of which should be executed independently from the other. To ensure independence, a regulatory agency separate from the European Commission has been established, the European Food Safety Authority (EFSA) (EC Regulation No. 178/2002, Ch. III). No administrative body has authority over the EFSA. The Management Board is composed of one representative of the Commission and 14 members who are appointed by the Council in consultation with the European Parliament (EP) from a list drawn up by the Commission. The 14 members are selected on the basis of their competence and expertise. The principle of the broadest possible geographic distribution within the

Union comes second to their personal qualities. Four of the members must have a background in organizations representing consumers and other interests in the food chain (Art. 25). The members have a term of office of four years that can be renewed once. The EFSA's scientific work is achieved by a scientific committee and scientific panels. These are composed of independent scientific experts. While the EFSA is responsible for risk assessment, the European Commission retains the initiative in areas of risk management and risk communication.

The principle of risk analysis requires that scientific information is available to decision-makers. However, this scientific information is by no means the only basis for the decisions to be taken. This follows from the definition of risk management (Art. 3.12): ' "risk management" means the process, distinct from risk assessment, of weighing policy alternatives in consultation with interested parties, considering risk assessment and *other legitimate factors*, and, if need be, selecting appropriate prevention and control options' (emphasis added). Furthermore, the precautionary principle articulated in Article 7 of the General Food Law creates the possibility of taking provisional risk management measures in cases of scientific uncertainty.

The procedural provisions in the specific regulations concerning GM food and feed are based on the so-called 'one door–one key' principle. This means several things. First, a single authorization for GM foods is valid throughout the Community (Art. 7.5). Unlike pharmaceutical products, there is no need to acquire authorization from each Member State where a product is brought onto the market. Second, the Regulation on GM Food and Feed allows for a single application to be filed for the purpose of obtaining the authorizations both under the Deliberate Release Directive, and the Regulation on GM Food and Feed (Arts. 5.5a and 7.8). This single application is followed by a single *risk assessment process* for which the EFSA is responsible, and a single *risk management process* (decision) involving both the Commission and the Member States through a regulatory committee procedure. Finally, if a product is likely to be used as both a food and a feed, it must be authorized for both purposes or not at all.

2.2.2 Application for authorization

In general, no pre-market approval of foods and food ingredients with a tradition of safe use within the EU is required. Products that are not normally consumed as food can be approved as additives. Foods that have no history of use in Europe before 1997 are considered novel, and require pre-market approval (in accordance with the Regulation on Novel Foods, 1997). GM foods are a special category of novel foods, and since 2004 are subject to a separate regulatory framework.

GM foods[5] require authorization (Regulation on GM Food and Feed, Art. 4.2) on the basis of a double safety assessment before they may be brought onto market. Authorization is required for the deliberate release of a GMO into the environment under the criteria laid down in Directive 2001/18 (Deliberate Release Directive), and likewise for the use of a GMO in food or feed under the criteria established in the Regulation on GM Food and Feed.

Although the role of the national authorities is limited, they can exercise some influence. They are no longer in charge of the procedure, but serve as box office to receive applications (Art. 5.2). In the procedure, Member States have the right to make comments or objections, and the EFSA can ask national authorities to carry out risk assessments (Art. 6.3b/c). Finally, national authorities can exercise some influence through the committee or through the Council.

Applications are to be submitted first to the competent authority of the Member State where the GM food product will be marketed first (Art. 5.2). The application must define the scope of the application, indicate which parts are confidential (Art. 30), and include a monitoring plan, a labelling proposal and a detection method for the new GM food or feed. The applicant must present available copies of studies that have been carried out and any other material demonstrating that the GM food complies with the criteria mentioned below (Art. 5.3). As environmental issues fall outside the scope of this chapter,[6] I will limit myself to the application of the Regulation on GM Food and Feed.

2.2.3 Processing the application

National authorities must acknowledge receipt of applications in writing within 14 days and inform the EFSA (OJ L 102/14, 2004, Art. 5.2a). Next, the application and any supplementary information supplied by the applicant must be made available to the EFSA, which is responsible for a scientific risk assessment covering both environmental risks and a human and animal health safety assessment. The EFSA's opinions are made available to the public, which is allowed to make comments.

The EFSA's assessments are subject to a six-month time limit, although this may be extended if the EFSA requests further information from the applicant (Art. 6.1). Within three months of receiving the EFSA's opinion, the Commission will draft a proposal for granting or refusing authorization on the basis of that opinion (Art. 7). The proposal must be approved by a qualified majority of the Member States within the Standing Committee on the Food Chain and Animal Health (Arts. 35 and 58), which is composed of representatives of the Member States. If the Committee issues a favourable opinion, the Commission adopts the

Decision. Alternatively, the draft Decision is submitted to the Council of Ministers for adoption or rejection by qualified majority. If the Council does not act within three months, the Commission adopts the decision. It is of course frustrating for applicants that no single body is responsible for the decision on applications. Instead, depending on opinions and coalitions, the power to decide moves up and down between the Commission, the Standing Committee and the Council. The GM Food and Feed Regulation has surprisingly little to say on the substance of the assessment that must be made in deciding the application. The Regulation stipulates that GM food and feed must not (Art. 4):

- have adverse effects on human health, animal health, or the environment;
- mislead consumers;
- differ from the food or feed it is intended to replace to such an extent that its normal consumption would be nutritionally disadvantageous for the consumer or animals.

Once granted, market authorizations for GM foods are valid for ten years throughout the Community (Art. 7.5). Conditions and restrictions may be connected to authorizations, and the applicant may be obliged to implement a post-market monitoring plan (Arts. 5.3k, 5.5b and 9). Products authorized must be entered into a public register of GM food and feed (Art. 28). Authorizations are renewable for ten-year periods (Art. 11), but are not untouchable during this period. The EFSA can, on its own initiative or on request, issue an opinion that an authorization for a product no longer meets the conditions. The Commission may then decide whether the authorization shall be modified, suspended or revoked.

2.2.4 Food and feed

Where a product is likely to be used as both a food and a feed,[7] a single application is submitted that gives rise to a single opinion from the EFSA and a single Community decision (Art. 27). This provision is a manifestation of the Commission's holistic approach, as typified by the ambition, under the new European food safety law, to encompass the whole food chain 'from farm to fork'. It means that applicants need not request separate authorizations for the use of the GMO in feed or food, but that a single risk assessment and a single authorization is undertaken for a GMO and all its possible uses. This is meant to ensure that the US accident with StarLink maize (a GM maize that, although it was only authorized for feed, also turned up in food) cannot occur within the EU: GMOs likely to be used as food and feed can only be authorized for both uses, or not at all.

2.2.5 Legal protection

Article 36 of the Regulation on GM Food and Feed states:

> Any decision taken under, or failure to exercise, the powers vested in the Authority by this Regulation may be reviewed by the Commission on its own initiative or in response to a request from a Member State or from any person directly and individually concerned.
>
> To this effect a request shall be submitted to the Commission within two months from the day on which the party concerned became aware of the act or omission in question.
>
> The Commission shall take a decision within two months requiring, if appropriate, the Authority to withdraw its decision or to remedy its failure to act.

Apart from what is enshrined in the Treaties and cases law from the European Court of Justice (ECJ), there exists no comprehensive body of administrative law in the EU. For this reason, special provisions concerning legal protection and other procedural matters have been elaborated on an ad hoc basis. The Treaties provide a legal basis for challenging decisions of the institutions, including the Commission, before the ECJ. However, this does not include decisions from independent agencies. It is therefore in accordance with to Article 47 of the General Food Law that the EFSA is brought under the jurisdiction of the ECJ for both contractual and non-contractual liability.

Against this backdrop, the review clause in Article 36 of the Regulation on GM Food and Feed is unnecessary. The Regulation does not grant the EFSA decision-making powers, and in this respect the clause is meaningless. It is positive that, in case of inaction on the part of the EFSA, the Commission can ensure the continuation of the authorization procedure by requesting a risk assessment from some other (national) authority. However, the downside of this provision is that it gives the Commission the power to put its own opinion in place of the EFSA's. This is a serious encroachment on the EFSA's role and the system for risk assessment more generally. It must be recalled that the separation of risk assessment from risk management was considered one of the leading principles of the new system of European food law in the General Food Law.

2.3 Labelling

One of the principles of modern European food law as laid down in the General Food Law is that of informed choice. To this effect Article 8 of the General Food Law states:

> Food law shall aim at the protection of the interests of consumers and shall provide a basis for consumers to make informed choices in relation to the foods they consume. It shall aim at the prevention of:

(a) fraudulent or deceptive practices;
(b) the adulteration of food; and
(c) any other practices which may mislead the consumer.

In relation to GM food, the principle of informed choice requires that consumers be informed of the use of gene technology in the production of the food products they buy or consume. In the EU, many consumers regard the use of gene technology as a relevant quality attribute of food products, mostly in a negative sense. The reasons for consumers' reservations as regards GM food are diverse, and include concerns about long-term safety. Article 13 of the Regulation on GM Food and Feed prescribes that the words 'genetically modified' or 'produced from genetically modified [name of the ingredient]' appear in the list of ingredients.[8] This requirement has been further elaborated in Article 4.6 of the Regulation on Traceability and Labelling:

> For products consisting of or containing GMOs, operators shall ensure that:
> (a) for pre-packaged products consisting of, or containing GMOs, the words 'This product contains genetically modified organisms' or 'This product contains genetically modified (name of organism(s))' appear on a label;
> (b) for non-pre-packaged products offered to the final consumer the words 'This product contains genetically modified organisms' or 'This product contains genetically modified (name of organism(s))' shall appear on, or in connection with, the display of the product.

These labelling requirements apply to all products derived from GMOs, even highly refined ones. An exception is no longer made for products in which no protein or DNA is present.[9] Indeed, there are very few exceptions. The labelling requirement does not apply to foods containing material that contains, consists of, or is produced from, GMOs in a proportion no higher than 0.9 per cent (Art. 12.2)[10] of the food ingredients considered individually or food consisting of a single ingredient, provided that this presence is adventitious or technically unavoidable.

The burden of proof is on industry: 'In order to establish that the presence of this material is adventitious or technically unavoidable, operators must be in a position to supply evidence to satisfy the competent authorities that they have taken appropriate steps to avoid the presence of such material' (Art 12.2 and Art. 47). To comply with this requirement food business operators must have detailed information at their disposal concerning the history of the raw materials they use. By way of example, research has shown that it takes a distance of 24.5 metres between a field of GM maize and a field of non-GM maize for the latter to remain below the 0.9 per cent threshold. Other products require even greater distances (Tweede Kamer 2004, p. 8). Labelling standards hence imply complete transparency of production chains and strict segregation of the entire GM food chain from the whole conventional food chain.

2.4 Handling

2.4.1 Monitoring

With the authorization, an obligation of post-market monitoring can be imposed. The authorization holder is obliged to report to the Commission on monitoring activities and a public version of the monitoring reports must be made accessible to the public (Art. 9.1). The authorization holder must inform the Commission of any new scientific or technical information that might influence the evaluation of the food's safe use.

2.4.2 Traceability

The General Food Law requires that food, feed, food-producing animals, and any other substance intended or expected to be incorporated into a food or feed must be traceable at all stages of production, processing and distribution (EC Directive No. 178/2002, Art. 18). To this end, food and feed business operators must be able to identify any person who has supplied them with a food, feed, food-producing animal, or any substance intended or expected to be incorporated into a food or feed. They must also be able to identify all businesses to which their products have been supplied. In other words, traceability is required one step up and one step down. Business operators must have systems and procedures in place that allow this information to be made available to competent authorities on demand.

Food or feed that is or is likely to be placed on the market in the Community must be adequately labelled or identified to facilitate its traceability through relevant documentation or information in accordance with the relevant requirements of more specific provisions. For GM food and feed more specific provisions apply. The Regulation on Traceability and Labelling defines 'traceability' as the ability to trace GMOs and products produced from GMOs at all stages of their production and distribution. Article 1 of the Regulation No. 1830/2003 (Regulation on Traceability and Labelling) states that traceability has as its objectives the facilitation of accurate labelling, monitoring the effects on the environment and, where appropriate, on health, and the implementation of the appropriate risk management measures including, if necessary, withdrawal of products. Under the General Food Law a food business operator is obliged to take unsafe food products off the market immediately.

Regulation 1830/2003 requires that specified information accompany GM food through all stages of the food chain.

The European Commission must devise a system of unique identifiers that are to be assigned to each GMO (Commission Regulation No. 65/2004). At the first stage of the placing on the market of a product consisting of or containing GMOs, including bulk quantities, operators shall ensure that the following information is transmitted in writing to the operator receiving the product:

- that it contains or consists of GMOs;
- the unique identifier(s) assigned to those GMOs.

Regulation No. 1830/2003 stipulates that at every following stage the same information must be passed on for each ingredient or additive that it concerns (Art. 4.1–4.2). This may seem straightforward, but in practice it is very difficult to preserve the identity of each raw material through to the end products in which they are used. Identity preservation (or internal traceability) is hard to realize, for instance in continuous production processes and in cases where failed products re-enter the production chain as raw materials. All information must be kept for five years (Art. 4.5 and Art. 5.2).

Small traces – not exceeding 0.9 per cent – are exempted from the traceability requirements if they are adventitious and unavoidable (Art. 7 amending Art. 21 of Directive 2001/18).

2.5 Emergency Measures

In cases where it is evident that authorized GM products are likely to constitute a serious risk to human health, animal health or the environment, the Commission can take the measures provided for in Articles 53 and 54 of the General Food Law (Art. 34). These measures concern the Regulation's new rapid alert system for feed and food (EC Regulation No. 178/2002, Arts. 50–52). The system foresees mandatory notification of any direct or indirect risk to human health, animal health or the environment within a network consisting of national competent authorities, the EFSA and the European Commission. It builds upon the former rapid alert system for food, extending it to include the feed sector and feed and food imports from outside the EU. The European Commission is entrusted with managing the system and ensuring immediate transmission of information to all contact points. Participation in the rapid alert system is in principle open to candidate countries, third countries and international organizations – subject to negotiated agreements. The EFSA's role is to supply scientific and technical information that will be helpful to Member States in deciding follow-up steps. If an alert is given through the network,

Member State authorities must take appropriate steps to inform the public that there are reasonable grounds to suspect a risk.

The General Food Law confers special powers on the European Commission for taking emergency measures. Such measures can be taken where it is evident that feed or food originating in the EU, or imported from a third country, is likely to constitute a serious risk to human health, animal health or the environment, and that such a risk cannot be satisfactorily contained by measures taken by the Member States. Such action can be initiated by the Commission itself or be requested by a Member State. Depending on the gravity of the situation, emergency measures can include the suspension of a feed or food from the market, or special conditions or other appropriate interim measures restricting the product's marketing or use.

3 NATIONAL REGULATIONS

The European framework does not fully pre-empt the powers and responsibilities of national authorities.

3.1 Enforcement

In general – and also in the case of GM foods – the enforcement of European law is the responsibility of the Member States. Article 17.2 of the General Food Law charges Member States with the responsibility to enforce food law through controls and sanctions. The Regulation on GM Food and Feed and the Regulation on Traceability and Labelling articulate this principle for GM food as follows:

> The Member States shall lay down the rules on penalties applicable to infringements of the provisions of this Regulation and shall take all measures necessary to ensure that they are implemented. The penalties provided for must be effective, proportionate and dissuasive. The Member States shall notify those provisions to the Commission six months after the date of entry into force of this Regulation at the latest and shall notify it without delay of any subsequent amendment affecting them. (Art. 45: almost identical is Art. 11)

and

> Member States shall ensure that inspections and other control measures including sample checks and testing (qualitative and quantitative), as appropriate, are carried out to ensure compliance with this Regulation. Inspection and control measures may also include inspection and control regarding the holding of a product. (Art. 9)

3.2 Coexistence and Segregation

The issue of coexistence refers to the ability of farmers to provide consumers with a choice between conventional, organic and GM products that comply with European labelling and purity standards.[11] European legislation has left this issue for Member States to regulate, and the Commission has issued guidelines to help focus national discussions (European Commission 2003). In brief, different situations are distinguished in which admixture can occur. Similar crops may become mixed between neighbouring fields but also over greater distances and over different seasons on the same field. All these possibilities require a different response. These can be taken on the farm (for example, isolation distances, buffer zones, pollen traps, etc.), in cooperation between farms (information exchange, agreement on product type, etc.), or at the level of public law.

However, it takes more than preventing admixture on the farm to ensure a level of purity that is in conformity with the thresholds. The whole logistic, production, packaging and distribution chain must be managed in such a way that the achieved level of purity is maintained. The required management measures can be termed segregation. Segregation must take into account the use and re-use of bags used to store cacao beans in Third World countries, the cleaning of trucks after transport, bulk transport by ship, storage in silos, continuous production processes, re-entry of faulty products in the production process, etc. So far, no guidelines on segregation have been issued. The Dutch government has opted for self-regulation as the best way to realize coexistence (Tweede Kamer 2004, Nos. 1–7).

3.3 (Product) Liability

Directive 2004/35/EC on Environmental Liability with Regard to the Prevention and Remedying of Environmental Damage was formally adopted on 21 April 2004 (Environmental Liability Directive). In principle, environmental damage covered by this Directive includes damage caused by the release of GMOs. However, the scope of application of this Directive is limited and, for numerous other reasons that cannot be explored here, the Directive is of limited use in respect of damage caused by GMOs (Wennerås 2005). This means that for issues of liability national (tort) law as yet is the most important source of law. As far as product liability law is concerned, national laws have been harmonized by the Product Liability Directive (Directive 85/374/EC Concerning Liability for Defective Products 1985).

The Regulation on GM Food and Feed explicitly states: 'The granting of authorization shall not lessen the general civil and criminal liability of any food operator in respect of the food concerned' (Art. 7.7). Nevertheless, safety assessments may greatly enhance the force of producers' 'development risk defence' under product liability law. This means that the producer escapes liability if the state of scientific and technological knowledge at the time when the product was put into circulation did not allow the existence of the defect to be discovered.

Quite another question is who is liable in case of cross-contamination[12] and who bears the costs incurred by the necessity to take segregation measures (Van der Meulen 2003, p. 4). With regard to the first question the Commission takes the position that the answer may differ from country to country, and that food business operators should be informed by the national authorities. In this respect, the Commission observes in its guidelines:

> The type of instruments adopted may have an impact on the application of national liability rules in the event of economic damage resulting from admixture. Member States are advised to examine their civil liability laws to find out whether the existing national laws offer sufficient and equal possibilities in this regard. Farmers, seed suppliers and other operators should be fully informed about the liability criteria that apply in their country in the case of damage caused by admixture.
>
> In this context, Member States may want to explore the feasibility and usefulness of adapting existing insurance schemes, or setting up new schemes.

With regard to the second question, it seems fair that those producers who want to continue to market conventional products are relieved of some of the administrative burdens necessary to separate traditional products from newer products. If, on the other hand, solutions take the form of compensation from public funds, questions concerning state aid or agricultural subsidies may arise.[13]

3.4 Intellectual Property Rights

To a certain extent, the field of intellectual property rights has remained national in character, although harmonization has taken place at the European and global levels, the latter notably through the TRIPs Agreement.

The infamous *Monsanto Canada Inc.* v. *Schmeiser*[14] case shows that cross-contamination in certain circumstances may trigger liability for infringement of IP rights, which is another reason why thorough measures to ensure coexistence through segregation seem vital. The case may be summarized as follows. Monsanto marketed seeds of different genetically modified crops. Crops – canola – were found on the field of Schmeiser with a genetic structure that matched the Monsanto product. Monsanto held

Schmeiser liable for infringement of patent rights. In defence, Schmeiser argued that this could be a case of spontaneous cross-breeding. Sowing seed might have dropped off passing trucks or perhaps pollen had drifted in by the wind. Schmeiser in turn held Monsanto liable for this contamination. Schmeiser lost in two instances. In practice this could mean that anyone who grows crops of genetically modified plants, whether or not intentionally, can be held liable, at any rate under Canadian law, to the patentee.

4 GLOBAL REGULATIONS

The reluctant European approach to GM food is considered by the US to constitute a barrier to trade. The question is raised why GM labelling should be required, particularly as consumers may perceive such labelling to constitute a warning. After a thorough safety assessment, it is argued, such a warning is inappropriate. Segregation and traceability require huge investments and make bulk transport and storage of products such as maize and soy virtually impossible. The US decided to take action on different fronts against these perceived barriers to trade.

In the spring of 2003 the US initiated legal proceedings with the WTO and public relations initiatives through the media. The publicity attack was opened by the American president in a speech delivered on 21 May 2003.[15] The President accused Europe of conducting a policy based on unfounded, unscientific fears. He added that this policy hindered the fight against hunger and poverty in Africa.[16] The accusation was inspired by countries like Zambia and Zimbabwe, which, despite imminent famine, were seriously considering refusing food aid that consisted of genetically modified corn from the United States. Moreover, Mozambique opposed the transport of this corn through its territory, mostly because it feared contamination of its own corn. Public health[17] concerns are at issue as well as concerns about the disappearance of outlet markets in the EU.[18] The latter concern could be realized if genetically modified corn is used as sowing seed or is disseminated spontaneously.

In the WTO litigation, the applicable legislative framework is the SPS Agreement, the WTO Agreement on Sanitary and Phytosanitary Measures (2003). The SPS Agreement starts from the premise that members have the right to take measures to protect the health of humans, animals and plants, even if these measures constitute a barrier to trade. The necessity of such measures must be established scientifically, be applied non-discriminatorily and may not constitute a disguised trade barrier. Sanitary or phytosanitary measures are deemed necessary if they are in conformity with international standards (Art. 3.2), in particular the standards, guidelines and recommendations of the Codex Alimentarius Commission (Annex 3a). If no international standard applies, or if a Member State desires to take more protective measures than those enshrined in the Codex, scientific proof of their necessity is required. By this provision the SPS Agreement adds legal significance to the Codex Alimentarius, which was not initially foreseen during the drafting of most of these standards.

One may wonder if the Codex Alimentarius has not been taken too lightly by the EU. The Codex seems to have been considered unproblematic because the EU safety and quality standards are usually higher than those of the Codex. Yet it is precisely these higher standards that may constitute barriers to international trade. This is illustrated by the beef hormones dispute that was triggered by an EU ban on US meat from cattle treated with hormones. The US filed a complaint with the WTO, and was found to be in the right.[19] The European authorities were unable to refute the evidence presented by the US that there were no risks. The example shows that the legality of the EU regulatory framework depends, amongst other things, on the outcomes of discussions that take place in the context of Codex Alimentarius and on the ability of the European authorities to substantiate their policy scientifically. The decision of the WTO arbitration panel in the GM case is expected in spring of 2006.[20] It is almost certain that the party found at fault will appeal.

5 CONCLUSIONS

In order to protect consumer safety and freedom of choice, EU legislators have created a complicated regulatory framework for GM foods, that places heavy administrative burdens on both food business operators dealing with GM foods and those who do not. Applicants for pre-market approval are confronted with shifting competences depending on opinions of the EFSA and the Commission and coalitions within the Standing Committee and the Council. In all stages of the food chain, food business operators who may come in touch with minute traces of GMOs must adhere to stricter labelling and traceability requirements than other business operators. National coexistence requirements place an even heavier burden on producers in the primary sector.

The investments necessary to comply with the new requirements stand the risk of having been in vain when the EU regulatory framework gives way under the pressure of international law and US opposition. It is evident that more thought needs to be put into designing a regime that fairly distributes the burdens associated with it.

NOTES

* The author wishes to thank Aaron Chase Underwood for his help.

1. The UK challenged this tendency to centralization in the case of the GMO regulations. The Commission chose Article 95 of the EC Treaty (harmonization) as their legal basis. Under Article 95 the European Parliament has co-decision and the Council can decide by qualified majority. The UK considered the creation of a central European approvals system not to be a harmonizing measure. Therefore the legal basis should in their view have been Article 308 of the EC Treaty that requires unanimity in the Council. However, in the legislative procedure the Commission got its way. So far the UK has not bring the case to the European Court of Justice.
2. Technically speaking the regulations entered into force the 20th day after publication (18 October 2003), but they 'apply' from six months after the date of publication.
3. The regulations refer to this situation as food produced 'from' GMOs.
4. This is consistent with the general observation that the instruments common in substantive food law consist of rules regarding the composition of food, rules regarding the handling of food and rules regarding the communication on food (Van der Meulen and Lugt 2004, pp. 65–83. See also Van der Meulen and Van der Velde 2004, pp. 145–6).
5. The Regulation on GM Food and Feed applies to GMOs for food use, food containing or consisting of GMOs and food produced from or containing ingredients produced from GMOs (Arts. 3 and 4.4).
6. It is interesting to note, however, that the reference made in the Regulation on GM Food and Feed to the Deliberate Release Directive implies a change in the status of the provisions involved from 'directive' to 'regulation'.
7. Provisions that apply to feed in particular fall outside the scope of this contribution. See Chapter III of the Regulation on GM Food and Feed (Arts. 15–26).
8. It should be noted that the word 'genetically' must appear on the label. This is because some additives are labelled 'modified starch', where the word 'modified' refers to conventional techniques of modification, not to gene technology.
9. However, processing aids fall outside the scope of the labelling requirements.
10. In case of a GMO that has not yet been authorized, a presence of 0.5 per cent maximum is considered not to constitute an infringement provided that this GMO has benefited from a favourable opinion from the Community Scientific Committee(s) or the Authority before the date of application of Regulation 2003/1829 (Art. 12.2 and Art. 47).
11. According to Regulation 2029/91 organic food has to be GMO-free. The detection limit is about 0.1 per cent.
12. For the question to what extent such a claim is possible under US law, see Rosso Grossman (2002) and (2003).
13. The first example of such a constellation is provided by the Danish Executive Order on Compensation for Losses due to Certain Occurrences of Genetically Modified Material. On 23 November 2005 the Commission authorized Denmark to pay compensation in cases where farmers with conventional or organic production suffer economic losses when genetically modified material is found in their crops. This was the first time that the Commission authorized such state aid. See IP/05/1458, EU Commission Press release, 23 November 2005.
14. See Federal Court Reports *Monsanto Inc.* v. *Schmeiser* (CA) [2003] 2 FC 165; http://decisions.fct-cf.gc.ca/fct/2002/2002fca309.html (appeal) and [2001] FCT 256; http://decisions.fct-cf.gc.ca/fct/2001/2001fct256.html (first instance).
15. NRC Handelsblad 22 May 2003. For the integral text see http://www.whitehouse.gov/news/releases/2003/05/20030521–2.html.
16. This accusation caused much irritation, especially since the EU spends seven times more on development aid than the US NRC Handelsblad 25 June 2003.
17. NRC Handelsblad 3 July 2002.
18. NRC 23 July 2002 and 2 August 2002.
19. The WTO's appellate body ruled on 16 January 1998. The ruling has the reference WT/DS48/AB/R and can be found at www.wto.org.
20. Final report (29 September 2006), see http://www.wto.org/english/news_e/news06_e/291R_e.htm.

REFERENCES

EC Directive 85/374/EC (Product Liability Directive) (1985), Concerning Liability for Defective Products, OJ L210/29.

EC Directive 2001/18/EC (Deliberate Release Directive) (2001), On the Deliberate Release into the Environment of Genetically Modified Organisms and Repealing Council Directive 90/220/EEC, OJ L 106/1.

EC Directive 2004/35/EC (Environmental Liability Directive) (2004), On Environmental Liability with Regard to the Prevention and Remedying of Environmental Damage, OJ L 143/56.

EC Regulation No. 258/97 (Regulation on Novel Foods) (1997), Concerning Novel Foods and Novel Food Ingredients, OJ L 43/1.

EC Regulation No. 178/2002 (General Food Regulation) (2002), Laying Down the General Principles and Requirements of Food Law, Establishing the European Food Safety Authority and Laying Down Procedures in Matters of Food Safety, OJ L 31/1.

EC Regulation No. 1829/2003 (Regulation on GM Food and Feed) (2003), Genetically Modified Food and Feed, OJ L 268/1.

EC Regulation No. 1830/2003 (Regulation on Traceability and Labelling) (2003), Concerning the Traceability and Labelling of Genetically Modified Organisms and the Traceability of Food and Feed Products from Genetically Modified Organisms and Amending Directive 2001/18/EC, OJ L 268/24.

European Commission (1999), *White Paper on Food Safety*, COM (1999) 719 final.

European Commission (2003), *Recommendation on Guidelines for the Development of National Strategies and Best Practices to Ensure Coexistence of Genetically Modified Crops with Conventional and Organic Farming*, OJ L 189/3, 2003.

Rosso Grossman, M. (2002), 'Biotechnology, Property Rights and the Environment', *The American Journal of Comparative Law*, **50**, 215–48.

Rosso Grossman, M. (2003), 'Genetically Modified Crops in the United States: Federal Regulation and State Tort Liability', *Environmental Law Review*, **5**, 86–108.

SPS Agreement (Agreement on the Application of Sanitary and Phytosanitary Measures) (1994), *International Legal Materials*, **33**, 1125–53.

Tweede Kamer der Staten-Generaal (2004), 'Coëxistentie Gg-gewassen, Conventionele en Biologische Gewassen, Verslag van een Algemeen Overleg', *Kamerstukken II*, 2003–2004, 29 404, Nos. 1–7.

Van der Meulen, Bernd M.J. (2003), 'Belastinggeld naar Gen-Voedsel?', *Staatscourant*, **247**.

Van der Meulen, Bernd M.J. (2004), 'Centralisatie van Voedselveiligheid in Europa', *Staatscourant*, **16**.

Van der Meulen, Bernd M.J. and M.J. Lugt (2004), 'Modern Europees Voedselveiligheidsrecht', *Justitiële Verkenningen*, **2**, 'Voedselveiligheid', 65–83.

Van der Meulen, Bernd M. J. and M. van der Velde (2005), *Food Safety Law in the European Union*, Wageningen: Wageningen Academic Publishers.

Wennerås, P. (2005), 'Permit Defences in Environmental Liability Regimes: subsidizing Environmental Damage in the EC?', *The European Yearbook of Environmental Law*, **4**, 149–80.

8. Restrictions on the cultivation of genetically modified organisms: issues of EC law*

Sara Poli

1 INTRODUCTION

Regional authorities have recently attempted to set up genetically modified organisms-free (GMO-free) areas, ban the cultivation of transgenic crops within their territories, or adopt measures to limit the cultivation of GMOs. Such measures are commonly referred to as 'coexistence measures' because they make coexistence between different forms of agriculture (conventional, organic and transgenic) possible.

This chapter explores EC law issues arising from the establishment of GMO-free areas and the adoption of coexistence measures. For this purpose, I will first examine whether the creation of GMO-free areas and the adoption of coexistence measures are compatible with EC law. Next, I will explore the division of competences between the Community and Member States in the adoption of measures restricting the cultivation of GMOs. It will then be possible to analyse the margin of manoeuvre left to Member States wishing to pose limits on the cultivation of GMOs.

These three issues are addressed in four sections. Section 2 describes the Commission's assessment of a legislative initiative effectively banning the cultivation of GMOs in Upper Austria (Land Oberösterreich), and addresses the compatibility of such policies with EC law. Section 3 focusses on the legal bases allowing the adoption of coexistence measures in EC law, and on the division of competences between Member States and the Community to deal with these measures. Section 4 provides an overview of draft legislation concerning coexistence put forward by several regional authorities in different Member States. Finally, Section 5 considers the powers retained by the Commission in the assessment of

* Section 2 of this chapter was originally published in the survey by Poli, S. and Notaro, N. (2005), 'Environmental Law 2002–2003', *Yearbook of European Law*, pp. 385–91, reproduced here by permission of Oxford University Press.

draft coexistence measures, and addresses both compatibility and competence issues.

2 GMO-FREE AREAS ASSESSED IN THE LIGHT OF ARTICLE 95(5) EC

In 2003 a few Austrian *Länder* sought to establish, by law, 'GMO-free areas' within their territory. Legislation was proposed in the *Länder* of Upper Austria, Salzburg and Carinthia.[1] Since legislative initiatives of this kind are capable of affecting the functioning of the internal market, they need to be notified to the Commission and assessed in the light of Article 95 EC.[2]

In 2003, draft legislation on 'Prohibiting Genetic Engineering' (hereinafter 'the draft legislation') by Upper Austria was examined by the Commission, under Article 95(5) EC and found to be contrary to EC law. The Commission issued Decision 653/2003/EC (hereinafter 'the Decision under examination'),[3] in which it sets out the reasons for its Decision.

Upper Austria resorted to Article 95(5) EC because it wished to adopt a new national measure, rather than maintain existing ones, with the aim of creating GMO-free areas. This contravened the framework set out in Directive 2001/18/EC (Deliberate Release Directive 2001),[4] which contains guiding principles pertaining to GMOs within the EU. For the purpose of putting the national measures into effect, the Austrian authorities notified the draft legislation to the Commission.[5] This was the first time that a Member State invoked Article 95(5) EC, also known as the 'Environmental Guarantee Clause',[6] to justify national legislation prohibiting the use of all GMOs.[7]

The primary aim of the draft legislation was to protect organic and traditional farming and animal products against 'contamination' by GMOs. It also aimed to protect biodiversity including natural genetic resources, especially in sensitive ecological areas whose existence could be threatened by the release of GMOs.

Substantively, the draft legislation sought to ban, for three years, the use of genetically modified seeds in Upper Austria.[8] It prohibited the experimental release of transgenic seeds, unless they were conducted in closed systems, and the use of transgenic animals for breeding and their release into the environment for hunting and fishing purposes. Adventitious traces of genetically modified seeds in conventional seeds were tolerated under a threshold of 0.1 per cent.

The Commission's reasoning that led to the rejection of the Austrian request for a derogation deserves a closer look. First, the Commission found that the draft notified departed from the Deliberate Release Directive on three counts. First, it banned all uses of GMOs in the territory of Upper Austria,[9] whereas the Deliberate Release Directive allows the release of GMOs into the environment on a case-by-case basis after individual risk assessments (Para. 52). Second, it required experimental releases of GMOs to be conducted in 'closed systems' (irrespective of any potential risk). Third, it set out a de minimis threshold (0.1 per cent) for the adventitious or technically unavoidable presence of non-authorized GMOs in seeds. However, as the Commission underlined, the Deliberate Release Directive does not have such a threshold. Member States therefore did not have discretion in judging which quantities of GMOs were dangerous and subsequently introducing such thresholds (Para. 55).[10]

The Commission went on to examine if the notified draft met the conditions to pave the way for authorization under Article 95(5) EC. The first condition is that 'new scientific evidence related to the protection of the environment or the working environment' must be presented that justifies the derogating measure. This evidence must be 'new' with respect to scientific information available at the time the Deliberate Release Directive was adopted. The second condition is proof of the existence of a problem 'specific' to the particular Member State.[11] The Commission found that none of these conditions had been satisfied. The 'new scientific evidence' provided by Austria consisted of the so-called 'Müller Study',[12] concerning coexistence of different forms of agriculture (organic, traditional and transgenic) released in April 2002, that is, after adoption of the Deliberate Release Directive. This study found that:

> If genetically modified varieties of seeds or planting material are cultivated extensively, genetically modified-free agricultural crop production would no longer be possible in future. . . . The same applies to transgenic animals used for breeding purposes and, to the release of transgenic animals especially for the purposes of hunting and fishing. In the long run, these animals reproduce and threaten the existence of the naturally occurring animal. (Para. 34, second subpara)

It should be noted that, whereas the economic consequences of the release of GMOs in Upper Austria are clearly identified by the Müller Study, the environmental implications of such releases received relatively little attention. This may be because the Müller Study underplayed environmental impacts or because as yet it remains difficult to prove environmental risks of transgenic crops.

Be this as it may, the Commission held that the Müller Study had not provided any new information that specifically concerned the environment (Para. 66). This turned out to be a decisive factor that gave rise to the rejection of the Austrian request for authorization.[13] The Commission claimed that the Müller Study was not 'new' because it contained 'data, which were for a large part available prior to the adoption of the framework Directive' (Para. 65). Although the study was released after adoption of the Deliberate Release Directive, the vast majority of the sources referred to in the bibliography predated publication of Deliberate Release Directive. This assessment was supported by the European Food Safety Authority (EFSA), whose opinion on the evidence provided by Austria was requested by the Commission.[14]

Although the Commission could have dismissed the Austrian request at this juncture, it went on to review the second condition: that the problem(s) arising from the release of GMOs should be specific to Upper Austria. The report accompanying the Austrian notification, drawing on the Müller Study, emphasized that the Austrian agricultural sector was based on small-structured farms and that in Upper Austria a high proportion of these farms produced organic crops. In order to safeguard production of organic agricultural products from GMO contamination, protection zones with a 4-kilometre radius from the foreign contamination source had to be created. However, the setting up of these areas would leave hardly any room available for GMO cultivation (see Paras. 35 and 36). The solution to this specific problem was the establishment of GMO-free zones.

This argument was rejected by the Commission. It considered small-scale farming not to be specific to Upper Austria, but something that existed in all Member States (Para. 70). The Commission's answer is not entirely satisfactory, since the specificity of the Austrian problem resided in the high number of small farms engaged in organic production, rather than the size of these farms.

Finally, the Commission considered whether the legislation notified was justified by the precautionary principle. This argument was dismissed because it was found to be too general and lacking in substance (Para. 73). The reason why reliance on the precautionary principle was essentially precluded was that the measure notified had been based, as the Court of First Instance (CFI) put it in *Alpharma*,[15] on 'a purely hypothetical approach to the risk, founded on mere conjecture, which has not yet been scientifically verified (Para. 156)'. Leaving aside the soundness of the Müller Study, it should be noted that, in any case, the prospect of Austria successfully relying on the precautionary principle as a justification to derogate from the Deliberate Release Directive was doubtful from the start. Following *Kingdom of Denmark* v. *Commission*,[16] it has become clear that the Commission contests Member States' right to invoke the principle in areas harmonized by EC law.[17]

The Commission concluded that, because Austria had failed to fulfil the conditions of Article 95(5) EC, its request for a derogation could not be accepted. The 'environmental guarantee' turned out to provide no basis to accommodate Austria's concerns about GMOs.

The outcome of the case is unsurprising and in line with previous decisions of the Commission in the context of Article 95(5) EC:[18] all decisions have been negative with only one exception.[19] The Commission's scrutiny of national measures affecting the internal market traditionally is very strict, especially when it evaluates new scientific evidence. Moreover, the environmental justifications in this particular instance lacked force, which made the outcome of the Commission's evaluation almost inevitable.

The Decision indicates that Member States enjoy very limited powers to depart from the regulatory framework pertaining to GMOs, in particular the Deliberate Release Directive. The latter, after all, has harmonized the environmental and health aspects associated with the release of GMOs into the environment, so that the reserved powers of Member States wishing to derogate from this Directive are limited. The Decision, which was subsequently challenged by Austria before the CFI,[20] sent a signal to other Member States wishing to enact legislation establishing GMO-free zones. It shows that it is hard to pursue policies of segregation in an internal market in which GMOs benefit from free circulation. Member States can only exclude specific GMOs, on a case-by-case basis, when they can prove that they are dangerous for human health or the environment. This does not mean that all aspects related to the cultivation of GMOs are to be decided at EC level merely by virtue of the fact that they may have an indirect impact on the internal market. As will be seen in the following section, factors such as farm structure, farming systems and the economic conditions of farmers in a given part of EC territory need be taken into consideration when making decisions on the cultivation of GMOs. This implies that national regional authorities enjoy certain powers in this area.

3 'COEXISTENCE MEASURE': BETWEEN COMMUNITY AND MEMBER STATES' COMPETENCE

Coexistence measures are aimed at avoiding the unintended presence of GMOs in conventional or organic crops, and can be defined as management measures adopted by farmers to avoid the admixture of genetically modified crops and non-transgenic crops. These measures enable farmers to engage in the three forms of agriculture with minimal mutual interference.

The adoption of these measures serves the interests of several different actors: national authorities wishing to safeguard the option of a non-transgenic agriculture, farmers of non-transgenic crops,[21] European consumers,[22] and farmers of genetically modified crops who otherwise would not have the option to choose for a transgenic agriculture.

The first problem to consider in relation to coexistence measures is *who* should develop these measures. Supranational, national and local authorities are all affected by coexistence measures. This is because such measures affect the functioning of the internal market, have an impact on (national) agricultural interests, and at the same time concern local authorities charged with their implementation.[23]

The issue of coexistence has been addressed by Community institutions for the first time in 2003. Therefore, a considerable time lapsed since enactment of the first Directives on GMOs in 1990. Article 26a of the Deliberate Release Directive, as amended by Regulation EC 1829/2003 on Genetically Modified Food and Feed (Regulation on GM Food and Feed),[24] is the first provision referring to the problem of admixture of GMOs and non-transgenic products, without explicitly mentioning the word 'coexistence'. This provision empowers Member States to adopt measures to avoid the unintended presence of GMOs in other products, whilst reserving a monitoring and advisory role to the Commission regarding the development of these measures.[25]

A few months after Deliberate Release Directive was amended, the European Commission issued a Recommendation on Guidelines for the Development of National Strategies and Best Practices to Ensure Coexistence of Genetically Modified Crops with Conventional and Organic Farming (Recommendation on Coexistence).[26] This Recommendation, which merely expresses the position of the Commission,[27] surprisingly does not contain any reference to Article 26a of the Deliberate Release Directive.

In this act, which is not legally binding, the Commission makes a few important statements. First, no form of agriculture, conventional, organic or genetically modified, should be excluded in the EU (Recital No. 1). Second, the issues of coexistence addressed in the Recommendation on Coexistence concern the 'potential *economic loss and impact* of the admixture of GM and non-GM crops, and the most appropriate management measures that can be taken to minimize admixture' (emphasis added, Recital No. 5). Third, the Commission considers that 'measures for coexistence should be developed and implemented by the Member States' (Recital No. 7). Fourth, the Commission's role is to support and advise Member States in this process by issuing guidelines for addressing coexistence (Recital No. 8). Such guidelines should provide a list of general principles and elements for the development of national strategies and best practices for coexistence (Recital No. 9).

The Recommendation on Coexistence is of importance also because in its recitals it clarifies issues of competence that Article 26a of the Deliberate

Release Directive had left unanswered. Member States are competent to adopt measures regulating economic aspects of the possible admixture of transgenic and non-transgenic crops. The Commission provides guidelines only on these aspects of coexistence.

However, not all issues pertaining to coexistence fall within Member States' competence. The Commission emphasizes the Community's competence to deal with the environmental and health-related aspects of coexistence.[28] The Community has exclusive competence in authorizing coexistence measures designed to protect *the environment and human health* from risks posed by GMOs because these aspects of coexistence are regulated by the Deliberate Release Directive, which is based on Article 95 EC. The implication is that only the Community institutions can address these aspects of coexistence in the framework of the authorization procedure to place a GMO on the market. Hence, with the adoption of the Deliberate Release Directive, Member States ceded to the Community the power to lay down the coexistence measures that a notifier[29] should comply with so as to protect the environment or human health from risks posed by the GMO concerned.[30]

The Commission's decision to leave the *economic aspects* of coexistence measures to Member States is inspired by the subsidiarity principle. National and regional authorities are better placed than Community institutions to adopt coexistence measures because these measures need to be adjusted to the specific geographic and environmental features of each Member State and region.[31]

The Commission's demarcation line between Community and Member States' competences, in as far as coexistence is concerned, is controversial. It leaves uncertainties as to where the precise respective limits of these competences are located. The separation between different aspects of coexistence seems artificial. It is doubtful whether in practice environmental and health-related aspects of coexistence can be separated from its economic aspects, thus giving rise to confusion as to who is competent to do what.

On the competence issue, the opinion of the European Parliament Committee on Agricultural and Rural Development proposes a different solution from that envisaged by the Commission. It calls for 'uniform and binding rules [on coexistence] to be established without delay at EC level',[32] without clarifying the role that national and regional authorities should play in the development of these rules. The Committee's remark can be interpreted as a preference for centralization of these measures at EC level. In the author's opinion, this preference is incompatible with the subsidiarity principle.

The Recommendation on Coexistence lists the general principles and factors[33] that national or regional authorities are advised to follow when elaborating coexistence measures. It is said that these measures, of which an indicative catalogue is provided,[34] should be scientifically based (Point 2.1.2). Concretely, they ought to take into consideration scientific evidence regarding the probability of admixture of GM and non-GM crops. They should also be taken at the appropriate level, which may mean at regional instead of national level so that they are appropriate for the conditions in which they come to apply (Point 2.1.6). Finally, they should be proportionate and not involve unnecessary burdens for farmers, seed producers and cooperatives (Point 2.1.4).

The Commission does not recommend any specific policy instrument to operationalize the coexistence strategy. It only draws attention to the fact that the reparation of economic damage caused by admixture of GM and non-GM crops requires the application of national civil liability rules, which should be amended to cater for this novel form of damage. The Commission also invites Member States to reflect on the suitability of existing insurance schemes to cover economic damage due to 'GMO contamination', and to set up new insurance schemes, should this be needed (Point 2.1.9).

4 THE DRAFT MEASURES ON COEXISTENCE

The Commission's refusal to authorize the setting up of GMO-free areas in Upper Austria had important repercussions for the strategy chosen by this region, other *Länder* and Member States that intended to protect conventional and organic agriculture from GMO contamination. Since obtaining a derogation allowing the setting up of GMO-free areas turned out to be difficult in areas harmonized by the EC legislation, the adoption of coexistence measures was considered an alternative solution to the creation of GMO-free areas. National legislation in this field seemed to stand a better chance of being considered compatible with Community law, since Article 26a of the Deliberate Release Directive could provide an ad hoc legal basis for measures intended to avoid the unintended presence of GMOs in other products. Moreover, as the Commission pointed out in its Recommendation on Coexistence, measures designed to deal with the economic consequences of coexistence are to be adopted at national level.

Following the rejection of Upper Austria's request for a derogation, many Austrian *Länder*[35] drafted coexistence measures while some Member States, such as Germany,[36] Denmark, Luxembourg and Italy,[37] adopted framework legislation in which they regulated the cultivation of GMOs, and announced that detailed measures on coexistence would be adopted in the future.

It is worthwhile looking at the content of these draft measures,[38] which make full use of the range of measures indicated by the Commission in its

Recommendation on Coexistence. These drafts, of which only an overview can be given here, have many features in common.[39] They invariably make explicit reference to the Recommendation on Coexistence or Article 26a of the Deliberate Release Directive. Their purpose is threefold: the prevention of the unintended presence of GMOs in other products, safeguarding of the option of a non-transgenic agriculture, and the protection of wild animals and plant species and their natural habitats, especially in protected areas or national parks.[40] They concern the cultivation of genetically modified products, which, generally speaking, are not banned but are placed under a strict regulatory control.[41] For example, landowners of genetically modified crops need an authorization[42] for the cultivation of transgenic crops,[43] have special notification and information duties with respect to their neighbours and the regional authorities, are subject to inspections and, in case of damage, have restoration duties. Special civil liability rules to compensate farmers who suffered economic loss as a result of admixture of genetically modified crops and ordinary crops are also provided for. It should also be noted that some draft legislation insurance cover for potential economic loss due to admixture is compulsory or a precondition for authorization.[44]

Some general observations on this draft legislation can be made. First, they are clearly inspired by the Recommendation on Coexistence. They contain the measures put forward in the Recommendation and include the controversial aspects such as civil liability for economic damage to which the Commission had drawn attention. However, they depart from the Recommendation on Coexistence in as far as they are not strictly limited to the economic aspects of coexistence but also concern environmental aspects of coexistence. For example, they make it more difficult or impossible to grow GMOs in environmentally sensitive areas. This means that the Community competence over the environmentally related aspects of coexistence, as defined in the Recommendation on Coexistence, is affected by these draft measures. Finally, they make coexistence possible, as indicated in the Recommendation, but under very strict conditions. Therefore, the suspicion arises that these draft legislations unduly restrict the cultivation of GMOs and could breach the provisions of the Deliberate Release Directive.

Everything considered, the Commission may be expected to address coexistence measures despite its acknowledgement that these measures fall, at least in part, within national competence. The Commission's right to have a say on coexistence measures has a legal basis not only in the Recommendation on Coexistence, but also in Directive 98/34/EC Laying Down a Procedure for the Provision of Information in the Field of Technical Standards and Regulations (Technical Regulations Directive),[45] which will be explored further in the next section.

5 THE COMMISSION'S POWERS WITH RESPECT TO DRAFT MEASURES ON COEXISTENCE

The Technical Regulations Directive concerns, amongst others, technical regulations that impact on trade. Each time Member States wish to adopt a technical regulation,[46] which in itself is capable of representing an obstacle to trade, they must notify the draft technical regulation to the Commission (Art. 8).[47] The Commission and the other Member States will make comments that will have to be taken into account as far as possible in the subsequent preparation of the technical regulation. The definitive adoption of the draft measure will be deferred for three months from the notification day (Art. 9(1)). The final adoption is postponed for six months if the Commission or the other Member States deliver a detailed opinion[48] 'to the effect that the measure envisaged may create obstacles to the free movement of goods' (Art. 9(2)). Coexistence measures fall within the definition of technical regulations, and therefore are subject to the notification and information procedure set up by the Technical Regulations Directive.[49] In compliance with the notification duties of the Directive, regional and national authorities sent all the draft measures mentioned in the previous section[50] to the Commission. The Commission's competent services delivered detailed opinions on these drafts, casting doubts on their compatibility with EC secondary legislation on GMOs.[51] As a result, the notifying authorities introduced changes in their draft measures. However, to the author's knowledge, none of these draft measures were adopted in their definitive form by 2005. (This information was kindly provided by a civil servant of the Commission, DG trade.)

The need to pass the Commission's scrutiny under the Technical Regulations Directive has become increasingly pressing for some Member States after the Commission decided to register several varieties of an approved genetically modified maize[52] in the common catalogue of varieties in September 2004.[53] The implication of this decision is that farmers who want to grow the authorized GMOs throughout Europe will be able to do so from now onward. Denmark, Italy, Greece, Germany and Poland showed uneasiness toward the Commission's move. In particular, they complained that such a decision was made before publication of a report on Member States' implementation of the rules governing coexistence.[54] These Member States also invited the Commission to set up a task force on coexistence so as to make sure that the collection and dissemination of information is coordinated.[55]

It may be wondered what happens if the Commission disapproves the draft measures on coexistence, but the notifying authorities nonetheless decide to adopt them. Having described the content of the draft legislation on coexistence, it is clear that the Commission would have good arguments to support its position that certain provisions of the draft legislation are incompatible with the Deliberate Release Directive. Member States could counterargue that the Commission's disapproval of the draft measures on coexistence encroaches upon Member States' competence, which is explicitly recognized in Article 26a of the Deliberate Release Directive and the Recommendation on Coexistence. The Commission could rebut this argument, pointing out that coexistence measures that have a direct bearing on the internal market fall within the powers of Member States, in as far as the Deliberate Release Directive is concerned with environmental and human health-related aspects of coexistence.

It is difficult to predict the outcome of a possible dispute before the Community Courts.[56] So far it seems that both the Commission and the notifying authorities are striving to reach a mutually acceptable compromise without resorting to litigation but this is no guarantee that litigation will be avoided in the future.

It is to be hoped that when assessing draft legislation in fields that have not been harmonized by EC law the Commission will show more leniency than with the examination of a national measure derogating from a harmonization measure under Article 95 EC. This would be of mutual advantage to the Commission and national authorities. The Commission could be satisfied because Member States would surrender the power to the enactment of radical measures such as the setting up of GMO-free areas and would give transgenic crops a 'chance' to co-exist together with other forms of agriculture. Member States would be able to decide how to use their territories without breaching the many obligations deriving from the free movement of goods provisions of the EC Treaty and the Deliberate Release Directive.

The legal problems mentioned in this chapter arose because there seem to exist an ontological difference of opinion between some Member States[57] and regional authorities that are contrary to the use of GMOs, and the Commission, which regards the use of biotechnology an as opportunity.[58] How may this divide be bridged? This is an open question. However, we can say that the awkward division of competences regarding powers to adopt coexistence measures articulated in the Recommendation on Coexistence does not help. As observed, the boundary between these competences is fuzzy, gives rise to confusion as to who is competent to do what, and contributes to tensions between these actors. In contrast to the Parliamentary Committee's report on coexistence, it is submitted that *all* aspects of coexistence measures should be left to national or regional regulators for the sake of legal certainty and in pursuit of the subsidiarity principle, whose operation was explicitly extended to regional authorities by the

Constitutional Treaty.[59] It goes without saying that, despite this re-allocation of competences, the Commission would still have a say on coexistence measures, should they boil down to disguised bans on the cultivation of GMOs. Moreover, it should be emphasized that this re-allocation could not cancel out the existing dichotomy between the Commission's approach to GMOs and that of some national or regional authorities.

6 CONCLUSION

The rejection of Upper Austria's request for derogation shows that GMO-free areas do not stand a great chance of finding the Commission's support. The reason why the Commission is strict in its evaluation of these initiatives is that they affect the internal market, and concern an area where harmonizing legislation has been adopted. Member States are allowed to adopt coexistence measures that restrict the cultivation of GMOs without banning them. This is an area in which there is no harmonizing legislation, and in which Member States' powers were recognized in secondary EC legislation. In addition, in the Recommendation on Coexistence the Commission acknowledged that these measures should be implemented by national or regional authorities, in line with the subsidiarity principle. In the same Recommendation, the Commission also emphasized the Community's competence over specific aspects of coexistence, regulated by the Deliberate Release Directive. The division of competences foreseen in this Recommendation can be criticized because it lacks clarity. Leaving full responsibility for the adoption of coexistence measures to national or regional authorities would have been preferable in the light of the subsidiarity principle. Such a solution would not prevent Commission control over these measures, because Member States are obliged to notify all draft technical regulations, including draft coexistence measures, pursuant to the Technical Regulations Directive. Therefore, should these measures be in contrast with the secondary EC legislation on GMOs or obstruct the free movement of goods, the Commission could ask the notifying authorities to change them.

At the time of writing, several draft measures on coexistence are being scrutinized by the Commission. Depending on the outcome of this process, coexistence measures could either turn out to be a catalyst for reconciliation between some national or regional authorities and the Commission, or another reason to restate their different approaches to GMOs.

Finally, another legislative development is worth mentioning in view of its links with the concept of coexistence promoted by the Commission. This is the proposal on the labelling threshold above which the presence of

GMOs in seeds should be indicated. The Commission has not yet adopted a formal proposal on this topic, despite at least two Member States having asked it to do so.[60] The future and the credibility of the coexistence policy of the Commission is linked to this initiative because, depending on the level at which the labelling threshold is set, coexistence of different forms of agriculture would be either encouraged or made virtually impossible.

NOTES

1. In this section only Upper Austria's draft legislation will be considered since only this region requested the Commission's authorization to enact this legislation on the basis of Art. 95(5) EC. The draft legislation proposed by the Austrian *Länder* of Salzburg and Carinthia can be found in the European Commission's TRIS (Technical Rules Information System database), notification No. 2003/475 and 2003/200.
2. This provision sets up a monitoring system intended to prevent the adoption of national legislation that would jeopardize the functioning of the internal market. This system is set in motion through a notification by a Member State, wishing to maintain or introduce national measures that derogate from a harmonization measure. The Commission has the power to authorize the enactment of the notified legislation only if it meets the conditions for derogation set out by Art. 95(4) or (5). For a comment on recent Commission Decisions based upon Art. 95 see Poli and Notaro (2004). On Art. 95 EC see De Sadeleer (2002) and (2003), Sevenster (2000) and Verheyen (2000).
3. (2003) OJ L 230/34.
4. (2001) OJ L 106/1. This Directive harmonizes, inter alia, the authorization procedure to place a GMO on the market. See Brosset (2004).
5. (2003) OJ C 126/4.
6. Paragraph 5 of Art. 95 EC was introduced in the Amsterdam Treaty. This paragraph allows a Member State 'to introduce national provisions based on new scientific evidence relating to the protection of the environment or the working environment on grounds of a problem specific to that Member State arising after the adoption of the harmonisation measure'.
7. The marketing authorization of a *single* (emphasis added) GMO can be suspended by national authorities within their territory on the basis of the safeguard clauses of the EC secondary legislation on GMOs. Art. 23 of the Deliberate Release Directive is an example, which is discussed below.
8. Essentially, the Austrian measure was intended to ban the cultivation of genetically modified seeds.
9. The Austrian ban concerned both GMOs already *authorized* to circulate within the internal market and future approvals of GMOs.
10. It should be noted that the Austrian and Danish delegations called upon the Commission to put forward as soon as possible a new proposal with respect to labelling of GMOs in seeds. The two delegations suggest that a labelling threshold should be set at a detection level of 0.1 per cent. See Council document No. 8689/04 of 21 April 2004, which is available in the Council's public register.
11. There are further conditions that a notified measure is required to meet in order to be authorized under Art. 95 EC. However, in the present case there was no need for the Commission to consider them because the Austrian draft legislation did not comply with the first two conditions.
12. This study was commissioned by Upper Austria and the Federal Ministry of Social Security, Generations and Consumer Protection.
13. As the Commission emphasized, 'the Austrian concerns about coexistence relate more

to a socioeconomic problem than to the protection of the environment or the working environment'. See Para. 67 of the Decision. The finding that the environmental justification for the Austrian ban was feeble was corroborated by the EFSA (European Food Safety Authority).

14. It was the first time that the Commission asked for the EFSA's opinion in order to assess requests for derogations under Art. 95(5) EC since the Authority had only just been established. Scientific evidence was assessed by the Scientific Committee on Toxicity, Eco-toxicity and the Environment in previous requests for derogations. However, this Committee will be replaced by the EFSA Committee as soon as its term of office comes to an end. See Commission Decision 2004/210, OJ (2004) L 66/45.
15. Case T-13/99 and T-70/99 [2002] ECR II-3495.
16. Case C-3/00 *Kingdom of Denmark* v. *Commission* [2003] ECR I-2643. For a comment on this case see Zander (2004).
17. Ibid, Para. 103. It should be noted that the ECJ has not clarified whether or not Member States are precluded from invoking the precautionary principle in areas harmonized by EC law.
18. See Dec. 2000/509/EC (2000) OJ L205/7 and Dec. 2001/570/EC (2001) OJ L202/37; Dec. 2002/59/EC (2002) OJ L 23/37; Dec. 2003/1/EC (2003) OJ L1/72.
19. This is Decision 2002/59, see above.
20. CFI (2005) Case T-235/04, *Republic* of *Austria* v. *Commission*, pending.
21. This is because in the absence of coexistence measures, were traces of GMOs to be found in non-transgenic crops, the producers of non-transgenic crops would suffer economic loss because they could not sell their products as 'non-transgenic' products on the market. They would have to indicate the presence of GMOs in the labelling information of these products.
22. The majority of European consumers do not want to buy transgenic products and are hence likely to be in favour of strict coexistence rules. On the European public attitudes to GMOs, see Tsioumani (2004).
23. Transgenic and non-transgenic crops can be safely cultivated next to each other, depending on the size, the nature and the location of the parcels of land. These conditions vary from area to area within the same Member State. This is why coexistence measures need be implemented at regional level.
24. This Regulation concerns genetically modified food and feed, OJ (2003) L 268/1. Art. 43(2) of Reg. EC 1829/2003 amended the framework Directive and introduced a new provision, Art. 26a, discussed below. For a comment to the Reg. EC 1829/2003 see Christoforou (2004) and Poli (2004).
25. Art. 26a of the Deliberate Release Directive, entitled 'Measures to avoid the unintended presence of GMOs' provides that:

    1. Member States may take appropriate measures to avoid the unintended presence of GMOs in other products. 2. The Commission shall gather and coordinate information based on studies at Community and national level, observe the developments regarding coexistence in the Member States and, on the basis of the information and observations, develop guidelines on the coexistence of genetically modified, conventional and organic crops.

26. Commission Rec. 2003/556/EC (2003) OJ L 189/86.
27. For example, the Parliament Committee on Agriculture and Rural Development considered coexistence measures in a report of its own. See the Report on coexistence between genetically modified crops and conventional and organic crops, adopted by this committee on 4 December 2003, document No. A-0465/2003.
28. 'Specific coexistence measures to *protect the environment and human health* (emphasis added), if needed, are included in the final consent of the authorization procedure in accordance with Directive 2001/18 . . ., with a legal obligation for their implementation' (Recital No. 4).
29. The person who submits a notification to national competent authorities of a Member State; the notification provides information on a specific GMO in order for the national

authority to decide whether the concerned GMO can be placed on the market or not. See Art. 2.6 and 2.7 of the Deliberate Release Directive.

30. Member States retain the power to act against risks to human health and environment posed by a GMO only *after* a marketing authorization is granted, should scientific evidence that such risks exist become available. Member States retain the power to provisionally restrict or prohibit the sale/use of an *authorized* GMO as or in a product within their territory, should this GMO constitute a risk to human health or the environment, under the safeguard clause (Art. 23 of the Deliberate Release Directive, emphasis added). Moreover, a derogation from authorization decisions taken on the basis of any harmonization measure, including the Deliberate Release Directive, is possible under Article 95(5) EC. However, we have seen how strictly the conditions of Art. 95(5) EC were applied by the Commission in the evaluation of Upper Austria's request for derogation.
31. The Recommendation recalls that 'the conditions under which European farmers work are extremely diverse. Farm and field sizes, production systems, crop rotations and cropping patterns, as well as natural conditions, vary enormously across Europe. This variability needs to be taken into account when devising, implementing, monitoring and coordinating coexistence measures. The measures that are applied must be specific to the farm structures, farming systems, cropping patterns and natural conditions in a region'. See Point 1.4 of the Recommendation.
32. Report on coexistence, see above.
33. A number of factors to take into account in the development of national strategies and best practices for coexistence are listed under Point 2.2 of the Recommendation.
34. These measures are indicated in Point 3 of the Recommendation. They include farming measures such as isolation distances between GM and non-GM fields, transport measures designed to ensure physical segregation, cooperation between farms in neighbourhood, monitoring schemes, land registration, information exchange and so on.
35. These are the *Länder* of Salzburg, Carinthia, Vienna, Lower Austria, Burgenland and Tyrol.
36. Germany announced that it would adopt this legislation and set out the main tenets of it in Council document No. 6458/04 of 18 February 2004, available in the public register of the Council.
37. See Italian Law of 28 January 2005 No. 5 published in *Gazzetta Ufficiale Repubblica Italiana* No. 22 of 28 January 2005. This legislation has not been notified to the Commission.
38. For the description of the content, the author relies on the English translations of the draft legislations made available online by the European Commission's TRIS (Technical Rules Information System database). It goes without saying that these texts are not legally binding since they are provisional and subject to changes.
39. This overview concerns draft legislations proposed in Germany, Denmark, Luxembourg, Burgenland, Lower Austria, Tyrol and Vienna and only addresses the most important common elements to the draft legislation. The overview is based on the texts of the drafts notified by the concerned Member States or regional authorities to the Commission and does not take into consideration the amendments suggested by the Commission because these suggestions are not made public. See TRIS notification No. 2004/133 (Germany), notification No. 2004/393 (Denmark), notification No. 2004/426 (Luxemburg), notification No. 2004/0459 (Burgenland), notification No. 2004/311 (Tyrol), notification No. 2005/5 (Lower Austria), notification No. 2004/538 (Vienna).
40. In addition to these aims, the Italian Law states that it is intended to protect 'the quality and the specificity of the Italian agricultural production'. See Art. 1 of the Italian Law, above note 37. The translation is made by the author.
41. For example, the cultivation of GMOs is prohibited or needs special authorization in ecologically sensitive areas.
42. Sometimes this authorization is additional to the Community authorization.
43. This authorization is additional to the Community authorization granted under the Deliberate Release Directive.

44. This is the case for the draft legislation of Lower Austria, Burgenland and Luxembourg.
45. (1998) OJ L 204/37.
46. This is defined as: 'Technical specifications and other requirements, including the relevant administrative provisions, the observance of which is compulsory, de jure or de facto, in the case of marketing or use in a Member State or a major part thereof, as well as laws, regulations or administrative provisions of Member States, except those provided for in Article 10, prohibiting the manufacture, importation, marketing or use of a product' (Art. 1(9)).
47. Although this is not explicitly provided in the Directive, a draft technical regulation needs to be notified when it concerns an area not harmonized by EC law. The idea behind the notification duty is to avoid Member States adopting technical regulations so as to create obstacles to the proper functioning of the internal market.
48. Within three months from the notification day.
49. For the purposes of the Technical Regulations Directive, it is irrelevant that coexistence measures fall in part within the Member States' competences.
50. See above. These notifications took place between March 2003 and the present time. It should be noted that the Italian Law, above note 37, enacted in January 2005, was not notified to the Commission. The Italian Law can be considered a framework law since it states that detailed rules on coexistence will be adopted in a future decree law. See Art. 3. The Italian authorities commit themselves to notify to the Commission the future decree law on coexistence, as prescribed by the Technical Regulations Directive. It should be emphasized that the Italian framework law prohibits the cultivation of GMOs, except for those approved for research purposes, until coexistence rules will be adopted in the form of the mentioned decree law. See Art. 8, Para. 1.
51. Information from the Commission.
52. This is the Monsanto MON 810 maize.
53. For this news see *BRIDGES Trade BioRes* **4**(16), 10 September 2004, published on the internet at http://www.ictsd.org/biores/index.htm.
54. The Commission had committed itself to write this report after two years from the adoption of the Recommendation on coexistence, that is to say in 2005. See Recital No. 10. Probably, the reason why the Commission could not postpone the authorization of the concerned GMOs is linked to the fact that the authorization procedure for the approval of the 17 varieties of genetically modified maize had to be completed within the time frame indicated by the Deliberate Release Directive. Any postponement not strictly related to the safety of the concerned GMOs for human health or the environment could be considered as unjustified.
55. See the information note of the Danish delegation to the Council, document No. 12834/04 of 28 September 2004, available online in the Council register.
56. The parties of this theoretical dispute would be, on the one hand, the notifying regional authorities, and on the other, the Commission. The former could challenge the legality of the Commission's disapproval of coexistence measure or the Commission's continued deferral of the date for the final adoption of the draft measure. It is assumed that these two measures are 'reviewable' acts within the meaning of Art. 230 EC.
57. Since the end of the 1990s, GMOs have become a cause of concern for a few EC Members who, in contrast with the provisions of the EC secondary legislation, opposed the use of GMOs as such or in food/feed products.
58. For an overview of the positive approach taken by the Commission in respect of biotechnology see the Commission's Communication 2002.
59. Art. I–11, Para. 3 of the Constitutional Treaty affirms that the Union takes action only if the objectives of this action cannot be achieved at central, *regional* and *local* (emphasis added) level.
60. See above.

REFERENCES

Alpharma v. *Council of the European Union* (2002), case T-13/99 and T-70/99, [2002] ECR II-3495.
Brosset, E. (2004), 'The Prior Authorisation Procedure Adopted for the Deliberate Release into the Environment of Genetically Modified Organisms: the Complexities of Balancing Community and National Competencies', *Journal of European Law*, **5**(10), 555–79.
Commission's Communication (2002), *Life Sciences and Biotechnology – a Strategy for Europe*, COM 27.
Court of First Instance (CFI) (2005), T-235/04, *Republic* of *Austria* v. *Commission*, pending.
Christoforou, T. (2004), 'The Regulation of Genetically Modified Organisms in the European Union: The interplay of Science, Law and Politics', *Common Market Law Review*, **41**(3), 637–709.
De Sadeleer, N. (2002), 'Les Clauses de Sauvegarde Prévues à l'Article 95 du Traité CE', *Revue de Tribunaux de Droit Européen*, **38**, 54.
De Sadeleer, N. (2003), 'Procedures for Derogations from the Principle of Approximation of Laws underArticle 95 EC', *Common Market Law Review*, **40**(4), 889–915.
EC Decision 2000/509 (2000), Concerning the Draft National Provisions Notified by the Kingdom of Belgium Concerning the Limitations of the Marketing and Use of Organostannic Compounds, *Official Journal*, L 205/7, 25 July 2000.
EC Decision 2001/570 (2001), On Draft National Provisions Notified by the Federal Republic of Germany on Limitations on the Marketing and Use of Organostannic Compounds, *Official Journal*, L 202/37, 13 July 2001.
EC Decision 2002/59 (2002), Concerning Draft National Provisions Notified by the Kingdom of the Netherlands under Article 95(5) of the EC Treaty on Limitations on the Marketing and Use of Creosote-treated Wood, *Official Journal*, L 23/37, 23 January 2002.
EC Decision 2003/1 (2003), Relating to National Provisions on Limiting the Importation and Placement on the Market of Certain NK Fertilisers of High Nitrogen Content and Containing Chlorine Notified by France Pursuant to Article 95(5) of the EC Treaty (notified under document No. C(2002) 5113), *Official Journal*, L 1/72, 18 December 2002.
EC Decision 653/2003 (2003), Relating to National Provisions on Banning the Use of Genetically Modified Organisms in the Region of Upper Austria Notified by the Republic of Austria Pursuant to Article 95(5) of the EC Treaty Decision under examination, *Official Journal*, L 230/34, 16 September 2003.
EC Decision 2004/210 (2004), Setting up Scientific Committees in the Field of Consumer Safety, Public Health and the Environment, *Official Journal*, L 66/45, 3 March 2004.
EC Directive 98/34 (1998), Laying Down a Procedure for the Provision of Information in the Field of Technical Standards and Regulations (Technical Regulations Directive), *Official Journal*, L 204/37.
EC Directive 2001/18 (2001) (Deliberate Release Directive), On the Deliberate Release into the Environment of Genetically Modified Organisms and Repealing Council Directive 90/220/EEC (Framework Directive), *Official Journal*, L 106/1, 12 March 2001.
Kingdom of Denmark v. *Commission* [2003], C-3/00, ECR I-2643.

EC Recommendation 2003/556 (2003), On Guidelines for the Development of National Strategies and Best Practices to Ensure the Coexistence of Genetically Modified Crops with Conventional and Organic Farming (the Recommendation on Coexistence), *Official Journal*, L 189/86, 23 July 2003.

EC Regulation 1829/2003 (2003), 'Concerning Genetically Modified Food and Feed', *Official Journal*, L 268/1.

Italian Law (2005) of 28 January 2005 No. 5, *Gazzetta Ufficiale Repubblica Italiana*, No. 22 of 28 January 2005.

Müller Study (2002), *On the Coexistence of Different Forms of Agriculture: Organic, Traditional and Transgenic*, released on April 2002, commissioned by the *Länder* of Upper Austria and the Federal Ministry of Social Security, Generations and Consumer Protection.

Notification according to Article 95(5) of the EC Treaty (2003), Request for Authorisation to Introduce National Provisions Incompatible with a Community Harmonisation Measure, *Official Journal*, C 126/4, 25 May 2003.

Parliament Committee (2003), *Report on Coexistence between Genetically Modified Crops and Conventional and Organic Crops*, adopted by this committee on 4 December 2003, document No. A-0465/2003.

Poli, S. (2004), 'The Overhaul of the European Legislation on GMOs, Genetically Modified Food and Feed: Mission Accomplished. What now?', *Maastricht Journal of European and Comparative Law*, **11**(1), 13–46.

Poli, Sara and Nicola Notaro (2004), 'Environmental Law (2002–2004)', in Piet Eeckhout and Takis Tridimas (eds), *Yearbook of European Law*, Oxford: Clarendon Press, pp. 629–74.

Sevenster, H. (2000), 'The Environmental Guarantee after Amsterdam: Does the Emperor Have New Clothes?', *Yearbook of European Environmental Law*, **I**, 291–310.

Tsioumani, E. (2004), 'Genetically Modified Organisms in the EU: Public Attitudes and Regulatory Developments', *Review of European Community and International and Environmental Law*, **13**(3), 279–88.

Verheyen, R. (2000), 'The Environmental Guarantee in Practice – a Critique', *Review of European Community and International Environmental Law*, **9**(2), 178–87.

Zander, J. (2004), 'The "Environmental Guarantee" in the EC Treaty: Two Recent Cases', *Journal of Environmental Law*, **1**(16), 72–9.

9. A tale of two commons: plant genetic resources and agricultural trade reform

Mary E. Footer*

1 INTRODUCTION

This chapter examines two different forms of commons (public goods enjoyed in common) in the field of agriculture and the intersection between them. One is the naturally occurring global commons of plant genetic resources for food and agriculture (PGRFA)[1] or crop germplasm. For generations this particular type of plant genetic resources has formed part of humanity's collective 'genetic estate' or common heritage of humankind, and has not been subject to individual appropriation. PGRFA were considered to be a freely available commodity, the only cost associated with their acquisition being that of collecting the germplasm. In practice, free availability translated into the free exchange paradigm that called for the unrestricted exchange of germplasm among farmers, plant breeders and scientists and is characteristic of a gift economy (Frow 1996, p. 94 and 1997, p. 198).

The global commons of PGRFA is currently under threat from the encroachment of private rights (Cullet 2001) in the form of individual property rights (both tangible and intangible) (Food Ethics Council 2002), upon public goods and institutions that are concerned with the conservation, use and management of those plant genetic resources. The process of encroachment manifests itself through various forms of appropriation including claims based on a growing number of provisions in major treaty instruments that straddle the fields of biodiversity, agriculture, food security, human rights and intellectual property protection. In short, we are witnessing the move towards 'privatization of the genetic commons'.[2]

The other commons is an anthropocentric localized commons of agriculture where the focus is upon the pursuit by national, regional and local governments of policy goals that may have no immediate economic value or where the prevailing policy goals rest on maintaining religious, symbolic or intrinsic values of importance to local communities and variably associated with the preservation of (agri)cultural heritage. The localized commons for agriculture may include policy goals that seek to preserve rural landscapes, traditional and indigenous farming practices or simply the rural way of life. Such societal goals, which encompass a broad range of cultural norms, are also at the heart of policies on environmental protection, food security and consumer concerns (health and food safety, animal welfare issues and the viability of rural areas).

At first sight this might not seem like a commons, in the traditional sense of public goods enjoyed in common,[3] but it covers a set of values and related societal goals that play a role in the economic life of rural areas. National, regional and local authorities, as represented by Member governments, are currently discussing this aspect of agriculture, as part of the agricultural reform process in the World Trade Organization (WTO). Some Members have chosen to characterize those concerns as positive externalities and to label them as public goods, which are jointly produced alongside food and fibre (Anderson 2001, pp. 112–16). At the same time the widespread development of genetically modified organisms or GMOs and the resort to proprietary technologies,[4] based on major intellectual property rights instruments, are changing the face of agriculture in many communities worldwide.

The localized commons of non-trade concerns, in the context of agricultural trade reform is characterized by a particular form of appropriation. Global trade rules, including those on trade-related intellectual property protection, are the driving force behind a form of market enclosure that is encroaching upon a domain normally reserved for sovereign governments in determining the appropriate policy framework for the preservation and management of rural areas.

My aim is to map the boundaries of these two commons, by telling their stories, noting their common provenance, observing their differences and discussing their inter-relationship. The starting point for the enclosure of both these commons is the process of commodification and the encroachment of market forces. By commodification I mean the process whereby the commodity form is extended to an object that is produced for use and exchange. One theory of the commodification of real property arises from the enclosure movement in England that took place from the fifteenth to the nineteenth century.[5] It changed communal and customary notions of real property and led to a wholesale transformation of agrarian practices, as a result of which rights were assigned away from their users.

A similar process of enclosure currently informs the commodification of crop germplasm, by means of its privatization and removal from the genetic commons.[6] Similarly, commodification is one of the forces behind the liberal and expansionist multilateral trading system in our contemporary market-driven society where it comes into conflict with the promotion of other non-trade values, as we shall presently note in the case of agricultural trade reform.

I turn first to an examination of the global commons for PGRFA or crop germplasm, which is under severe strain from international treaty regimes that either articulate private rights and interests or seek to accommodate private rights with public goods and institutions that are concerned with the conservation, use and management thereof. Whatever the means and by whatever process they are employed the overall effect is one of enclosure through a process of commodification that leads to the appropriation of public domain rights.

This is followed by an examination of the enclosure of the localized commons of agriculture in relation to non-trade concerns, made operational through a broad array of international trade agreements, the best known of which are the Multilateral Trade Agreements, concluded at the end of the Uruguay Round, in particular the WTO Agreement on Agriculture.

2 THE GLOBAL COMMONS IN PGRFA OR CROP GERMPLASM

Since time immemorial people have traded seed for food and fibre. Many farmers consider that seeds have an intrinsic as well as monetary value, since they contain the genetic material, or germplasm, which form the basis for the diversity of those species that support agricultural production such as soil biota, pollinators, predators and those species in the wider environment that support diverse agro-ecosystems (agricultural, pastoral, forest or aquatic). It has been estimated that the principles of crop cultivation and livestock domestication have been developed from just eight to ten centres of origin and crop diversity to cover almost the entire globe.[7]

Today the production of non-native plant genetic resources for food and agriculture, or PGRFA, is at the basis of agricultural production in every country (Ten Kate and Laird 1999, p. 118 and Thrupp 2000). Traditionally, most farmers have planted back their harvested seeds and improved basic crops such as rice, maize, millet and wheat without asserting individual or collective ownership over them (Singh Nijar 1999). In terms of both food production and crop development countries remain dependent on access to each other's basic crop germplasm in order to ensure food security for their citizens.

2.1 Evolution of the Global Commons in PGRFA

In the context of explaining the current boundaries of crop germplasm, it is necessary to understand some of the historical background, the key to which is the activity of plant breeders rather than farmers. Plant breeders have a wealth of national crop PGRFA at their disposal, boosted by access to foreign crop germplasm from public and private collections in their own countries, held in National Agricultural Research Centres (NARCs) and genebanks. In the late 1960s, access to germplasm from abroad increased for breeders when a series of 15 International Agricultural Research Centres (IARCs) were established across the globe, based in developing countries. Each IARC was given the task of improving a particular set of crops in a particular region. In 1971, the Consultative Group on International Agricultural Research (CGIAR),[8] led by the Rockefeller and Ford Foundations, was established to coordinate and support the activities of the IARCs.

As a result, over 3000 varieties of essential crop germplasm have been brought within ex situ[9] gene bank collections,[10] some of which are administered by scientific research stations as part of the CGIAR. The purpose of the CGIAR-administered international agricultural research programme has been to develop and promote science-based solutions to problems arising from constraints on the production and sustainability of basic food crops. Today the CGIAR is the world's largest publicly coordinated effort that aims to addresses global food security issues and environmentally sustainable development (Thornström 2005, 21–2).

Many scientists and some governments have always considered this essential crop germplasm to be held in trust for the benefit of the international community, in particular, for developing countries, in what is a form of international stewardship. More particularly, it forms part of humanity's collective 'genetic estate' or common heritage of humankind and is therefore not subject to individual appropriation (Footer 2004 and Raustiala and Victor 2004, 284–8). The concept of non-appropriation is a common feature of both the common heritage and global commons status (Redgwell 1999, p. 130), while the distinctive feature of common heritage is its emphasis on international management and the control and sharing of resources.[11]

Thus, those ex situ collections, which belong to the CGIAR and similar gene banks, have operated from the basic premise that essential crop germplasm should be a freely accessible and exchangeable commodity (free in the sense of unrestricted not necessarily in the sense of unpaid). The only cost associated with acquisition of this type of public domain PGRFA is that of collecting the germplasm. In practice, free accessibility translates into the free exchange paradigm, which mandates the unrestricted exchange of banked germplasm among plant breeders and other scientists.

In its formative years, in the absence of detailed legal regulation, the CGIAR norm of free exchange was sufficient to maintain the relatively free international flow of plant genetic material stored in the gene banks at international agricultural research centres or IARCs around the world. The norm developed spontaneously,[12] drawing on the practice of free exchange in which the small community of research scientists on PGRFA participated and resulting in the free exchange paradigm for essential crop germplasm.

2.2 Enclosing the Global Commons in PGRFA

The situation held good for a couple of decades when scientific research was mostly publicly funded. In addition, and despite the lack of a specific obligation to share benefits, the boundaries of PGRFA were often criss-crossed by (in)formal networks of breeders, predominantly in the public sector, which continued to exchange 'elite' germplasm.[13] However, by the early 1980s there had appeared a number of competing claims to plant genetic resources, which came from the three directions.

The first marks the beginnings of the enclosure of the global commons in plant genetic resources. It comes from the claim by mostly developing countries in the southern hemisphere to permanent sovereignty over their natural resources.[14] The second is the claim by states in the post-Stockholm era to ownership access, and control of natural resources, including genetic resources within their national boundaries where those resources are located and the right to their conservation and management (Declaration of the United Nations Conference on the Human Environment, Stockholm, 5 June 1972). The third claim, predominantly from the private sector, is the extension of intellectual property rights protection to living forms, including plants, plant varieties and their genetic information.[15] In the face of these claims and the potential enclosing effect of each them in terms of the exercise of state sovereignty, jurisdictional competence and the encroachment of private property rights on the public domain, the food and agriculture community has struggled to retain the traditional interdependence that plant breeders and scientists have always relied upon for the accession of essential crop germplasm.

At the 22nd session of the Food and Agriculture Organization (FAO) Conference in November 1983,[16] FAO Members adopted the International Undertaking, which sought to maintain the exception to the principle that states have permanent sovereignty over their natural resources because the sub-species of genetic resources, by their very nature form part of the common heritage of [hu]mankind and should be freely accessible. Plant genetic resources were public goods of economic and/or social interest, particularly for agriculture, which should be 'explored, preserved, evaluated and made available for plant breeding and scientific purposes'. Since they belonged to the heritage of mankind, they should be available without restriction'.[17] Thus, the International Undertaking called for the conservation of both in situ and ex situ collections of plant genetic resources (FAO Resolution 8/83, 1983, Article 4:1 and 4:3) and encouraged governments to adhere to the paradigm of free accessibility and exchange of plant germplasm within their jurisdictions for the 'purposes of scientific research, plant breeding or genetic resource conservation' on a free of charge basis (FAO Resolution 8/83, 1983, Article 5), or on mutually agreed terms.

However, from its very inception the International Undertaking was not without its detractors, partly because it included farmers' landraces, other traditional varieties and wild varieties, which were freely available, together with cultivated varieties in current use and newly developed varieties, which were often the products of formal breeding and subject to plant breeders' rights (FAO Resolution 8/83, 1983, Article 2:1(a)). In other words, the International Undertaking placed *all* plant genetic resources on an equal footing, thereby implying free access to all plant genetic material, including essential crop germplasm (Ntambirweki 2001, 111).

Developed countries disliked its lack of an explicit reference to plant breeders' rights as provided for under national legislation, in conformity with one of the acts of the Union Internationale pour la Protection des Obtentions Végétales (UPOV) (International Union for the Protection of New Varieties of Plants 1972). Developing countries were cautious because they felt the scope of the International Undertaking failed to recognize that many plant varieties originate in primitive cultivars (or landraces) and have been genetically improved by traditional and indigenous farmers. They were therefore keen to see some recognition of the rights of farmers to save, use, exchange and sell seed, irrespective of their origin. This factor combined with the increase in the number of individuals from developed countries, who acquired access to such resources without paying any form of compensation to indigenous and local farmers, led many of those farmers to claim the right to be considered as the contributing force behind the conservation and improvement of such crop germplasm.

2.3 Completing the Enclosure of the Global Commons in PGRFA

The International Undertaking was subsequently 'amended' by means of three 'agreed interpretations', each of which can be read as a form of enclosure, that frames the global commons in PGRFA and lays it open to potential acts of appropriation. The first interpretation acknowledged plant breeders' rights (as provided for in one of the Acts of UPOV) as *not* incompatible with the International Undertaking (FAO Resolution C 4/89, 1989). Thus, states could restrict the free access to and exchange of genetic material, or germplasm, in order to comply with their international obligations and national implementing legislation, thereby dealing a blow to the free exchange paradigm.

The 'trade-off' for this first act of enclosure was another in the form of the second interpretation, which for the first time gave substance to the idea of farmers' rights as 'rights arising from the past, present and future contributions of farmers in conserving, improving, and making available plant genetic resources, particularly those in the centres of origin/diversity' (FAO Resolution 5/89, 1989). It also determined that 'these rights [were] vested in the International Community, as trustee for present and future generations of farmers', thereby introducing a form of international stewardship, with explicit reference to the notion of inter-generational equity (Footer 2004, pp. 439 and 456–60).

The third interpretation completes the enclosure process in recognizing that 'the concept of the common heritage of mankind, as applied in the International Undertaking on Plant Genetic Resources, is subject to the sovereignty of states over their plant genetic resources' (FAO Resolution 3/91, 1991), in conformity with the Convention on Biodiversity (CBD). At the same time it implicitly limited the scope of the free access provision, thereby excluding breeders' lines and farmers' breeding materials from its purview and giving tacit recognition to the commodification of essential crop germplasm.

Thus, while the International Undertaking set out with the bold intention of recognizing a global common in plant genetic resources, in fact it ended up laying itself open to competing claims over those resources from national governments and private operators as a result of the encroachment of the market on the global commons of PGRFA. The process of enclosure also reveals the failure of the international community, under the auspices of the FAO, to provide adequate guardianship of a global commons in crop germplasm. While the FAO did set up a Global System for the Conservation and Utilisation of Plant Genetic Resources for Food and Agriculture (FAO Resolution C 9/83, 1983), of which the International Undertaking forms a part, it failed to receive adequate financial support. More importantly, in terms of ensuring a stable basis for crop diversity, it failed to develop its own gene bank network for the collection of crop germplasm at a critical point in the late 1970s that might otherwise have ensured a stable basis for the future of crop diversity.

The story of the enclosure of the global commons in PGRFA does not end there. In 1992, Agenda 21 (Robinson 1993, Chapter 16) called for the strengthening of the FAO Global System and its harmonization with the CBD (Robinson 1993, Chapter 26.4 and FAO Resolution 7/93, 1993), which meant that something had to be done about ensuring access to and benefit-sharing of PGRFA, including those ex situ collections not acquired in accordance with the CBD and farmers' rights (Girsberger 1999). Subsequently, a fundamental revision of the International Undertaking proceeded at the FAO, following Resolution 3 of the 1992 Nairobi Conference (Nairobi Conference 1992), which addressed the inter-relationship between the CBD and the promotion of sustainable agriculture, and the FAO Conference Resolution of 1993 (FAO Resolution C 7/93, 1993; Ten Kate and Diaz 1997). This led to a series of negotiations under the auspices of the FAO Commission on Plant Genetic Resources for Food and Agriculture (CGRFA), which began in 1994 and continued until the adoption of the International Treaty on Plant Genetic Resources for Food and Agriculture in 2001 and its recent entry into force (FAO Resolution 3/01, 2001, hereinafter ITPGRFA or 'Seed Treaty').

2.4 The Future Regulation and Governance of PGRFA

The scope and coverage of this newest FAO Treaty differ markedly from those of the previous International Undertaking and reflect the overall orientation of the CBD with its emphasis on the permanent sovereignty of states over their genetic resources. Its main objectives are the conservation and sustainable use of PGRFA combined with facilitated access to a specified list of PGRFA and the sharing of benefits arising from the utilization of that accessed crop germplasm in the form of information exchange, technology transfer, capacity building and commercial development. The mechanism that the Seed Treaty employs for the regulation of such access and benefit-sharing is known as the 'Multilateral System of Access and Benefit-sharing', or simply the Multilateral System (FAO Resolution 3/01, 2001, Article 10).

The Multilateral System represents a policy reversal in the field of PGRFA, which has seen essential crop germplasm pass from the domain of common heritage of humankind to that of national sovereignty over biological resources. The idea of facilitated access to essential crop germplasm completes the enclosure of the global commons in PGRFA (Safrin 2004, pp. 644–52) by establishing the present and future boundaries for access and benefit-sharing of crop germplasm. It does so by continuing to recognize that PGRFA, as a sub-species of plant genetic resources, forms part of the common concern of humankind[18] and that *all* countries depend to some extent on essential crop germplasm that originated elsewhere (Footer 2004, pp. 444–56; Cf. Mgbeoji 2003). At the same time, Parties to the Seed Treaty retain full authority to regulate their own PGRFA but the Seed Treaty differs from the CBD in the exercise of permanent sovereignty over those resources because it places strict limitations on the ability of the Parties to restrict access from other states.

While the Seed Treaty is intended to cover all genetic material for food and agriculture, Parties are under a minimum obligation to guarantee access to the genetic material of 35 crop genera and 29 forage species that are in the public domain and are considered essential for global food security and human nutrition.[19] The listing in Annex I to the Treaty includes crops like rice, maize, wheat, cassava and potatoes, which are important to agrobiodiversity and food security; a notable omission is soybean, which China managed successfully to have excluded from the basic list of food crops (Footer 2004). In addition, the Multilateral System includes essential crop germplasm listed in Annex I, which is held in the ex situ collections of the international agricultural research centres (IARCs) that are members of the CGIAR and the genebank collections of other international institutions (FAO Resolution 3/01, 2001, Article 11.5 together with Article 15.1(a)), both of which continue to form part of the common heritage of humankind.

Despite its enclosing effect, the Seed Treaty is a last-ditch attempt to align the terms of access to crop germplasm with publicly funded research and development needs. Access is intended primarily for the purposes of conservation and sustainable use in connection with research, breeding and training for food and agriculture.[20] This means that access must be granted expeditiously (FAO Resolution 3/01, 2001, Article 12, Paragraph 3(a) (b)), free of charge (or at minimal cost), with available 'passport' data and any other associated available non-confidential descriptive information (FAO Resolution 3/01, 2001, Article 12, Paragraph 3(c)). There is also a standstill provision that gives a developer (breeder or farmer) in a country of origin of crop germplasm the right to delay access for a period during which that crop germplasm is under development at the time when the access is requested (FAO Resolution 3/01, 2001, Article 12, Paragraph 3(e)).

This facilitated but relatively 'free' access is offset by deliberate ambiguity in the Seed Treaty provision that deals with the issue of plant genetic information and technology, which is protected by intellectual property rights and confidentiality clauses and which divided developed and industrialized countries during negotiations.[21] The compromise that was worked out favours the maintenance of essential crop germplasm in the public domain by preventing recipients of PGRFA through the Multilateral System from claiming intellectual property rights that might otherwise limit facilitated access to the PGRFA, or their genetic parts or components, in the form received from the Multilateral System (FAO Resolution 3/01, 2001, Article 12, Paragraph 3(d)). However, where essential crop germplasm is already protected by intellectual property rights, or indeed any other property rights (including communal property rights and the rights of indigenous peoples to genetic resources in some countries), access can only take place in conformity with the Treaty or other statutory instruments regulating the type of property rights (FAO Resolution 3/01, 2001, Article 12, Paragraph 3(f)). A further point is that PGRFA accessed under the Multilateral System must be made available to other interested parties by the recipient under the conditions laid down by the Treaty (FAO Resolution 3/01, 2001, Article 12, Paragraph 3(g)). The inclusion of these provisions was hailed as a victory by scientists and research institutions from countries with strong, government-backed research and breeding capabilities as well as many developing countries.

The benefit-sharing provisions under the Seed Treaty form part of what some people have termed a 'grand bargain' (Ten Kate and Laird 2000, 244), which seeks to redress the asymmetry in plant genetic resources and bargaining power between the gene-rich South and the gene-hungry North, through a process of corrective equity.[22] As a starting point, the Seed Treaty takes the approach that all benefits arising from use, including commercial use, of PGRFA acquired through the Multilateral System are to be shared equitably through one of four mechanisms: (1) exchange of information (catalogues and inventories, information on technologies and the results of technical, scientific and socioeconomic research); (2) access to and transfer of technology; (3) capacity-building (the establishment of programmes for scientific and technical education and training in conservation and sustainable use of PGRFA); and (4) sharing of monetary and other benefits arising from commercialization (FAO Resolution 3/01, 2001, Article 13, Paragraph 2, Paragraphs (a) through (d)). These mechanisms must take account of the priorities in the FAO's rolling Global Plan of Action,[23] which depends on the effective implementation of the benefit-sharing provisions, and the funding strategy, which is in the process of being set up under the Treaty in the form of a Trust Account. The Parties still need to address the commercial aspects of benefit-sharing, that is, the sharing of monetary and other benefits, including those arising from the involvement of the private sector in developing countries in research and technology development.

At a practical level, a standard Material Transfer Agreement or MTA will be used to facilitate access to crop germplasm through the Multilateral System, details of which are being considered by an Expert Group on the Terms of the Standard Material Transfer Agreement.[24] In its recent report the Expert Group recommended that a Contact Group be established by the Interim Committee for the ITPGRFA in order to draft the standard MTA, which needs to include specific provisions that make it clear who benefits from the Multilateral System, and how the providers of material can receive an equitable share of the benefits arising from the commercialization of products incorporating material accessed through the Multilateral System.

A further issue that the standard MTA must address is the manner in which recipients of the material will make future payments into the Trust Account in conforming with the Treaty, and whether some categories of recipients such as small farmers in developing and transition economy countries should be excluded from payments altogether.[25] It falls to the Governing Body – the key institutional body under the Seed Treaty's governance structure – to set the level, form and manner of payment under the commercialization provisions of Article 13.2(d) of the Treaty, in line with commercial practice, although the matter has already attracted considerable attention, as is clear from the Report of the Expert Group on the standard MTA (see FAO Resolution 3/01, 2001, Article 13, Paragraph 2(d)(ii), second full paragraph; Report of the Expert Group 2004, Paragraphs 25–32. See also Koo et al. 2003).

For the countries that are Parties to the Seed Treaty and for the crops that are covered under the Multilateral System, the Treaty to some extent re-establishes the idea that PGRFA are a public good, freely accessible and exchangeable for conservation and use by all but it does not re-establish the commons in PGRFA. While the concept of the common heritage of humankind, which was central to the International Undertaking, may have been abandoned in favour of a 'common concern of humankind' it may live on in Article 15 of the Treaty. In accordance with this Article those ex situ collections of crop germplasm that are held 'in trust' by the IARCs and similar institutions, which appear on the Annex I list and which were collected *before* the date of entry into force of the Seed Treaty (FAO Resolution 3/01, 2001, Article 15, Paragraph 1), will henceforth fall under the Multilateral System and will be subject to the Seed Treaty provisions.[26] They will be guided by a system of agreements to be concluded between them and the Governing Body under the Treaty.

Financial benefits flowing from the exchange of germplasm in the ex situ collections and falling under the Multilateral System, that is, the Annex I listings held by the IARCs and other seed banks, will accrue to the funding mechanism to be set up under the Treaty (the Trust Account), whereby such financial benefits will be applied to the conservation and sustainable use of the PGRFA in national and regional programmes located in developing countries, which are centres of genetic diversity, and least developed countries (FAO Resolution 3/01, 2001, Article 15, Paragraph 1(b)(iii)).

Crop germplasm in the ex situ collections, which is *not* on the Annex I list and was collected *after* the date of entry into force of the Seed Treaty will be made available on a bilateral basis, on terms mutually agreed between recipient IARCs and countries of origin, that is, those countries that possess genetic resources in situ and those countries that have acquired genetic resources in accordance with the CBD (both considered as source countries under the Seed Treaty) (FAO Resolution 3/01, 2001, Article 15, Paragraph 3).

3 THE LOCALIZED COMMONS IN AGRICULTURE

In some respects I think it is easier to portray the localized commons in agriculture, specifically when dealing with non-trade concerns. This is because agriculture can play several different roles in society in addition to its primary function of supplying food and fibre. It is well-known that various forms of agricultural activity can 'shape the landscape, provide environmental benefits such as land conservation . . . [ensure] sustainable management of renewable natural resources and the preservation of biodiversity' (Sensi and Werksman 2001). Of course agricultural activity can also bring about ground water contamination, threaten wildlife, introduce harmful pesticides and fertilizers into the environment, bring about soil erosion and lead to a loss of biodiversity through the use of monocultures.

Until the early part of the last century, farming was 'the principal occupation and lifestyle of most people'. Indeed, this is still true in many parts of the developing world where one half of the world's population of about five billion still derive their livelihoods from farming (Echols 2002, p. 29). Today, many small farmers and rural communities in developing countries continue to pursue a tradition that allows for diversified and heterogeneous agricultural production, relying on a mixture of wild species (landraces) and domesticated species of crop germplasm to ensure agricultural biodiversity.

This group of farmers also includes indigenous peoples and local communities, some of whom see their access to agricultural biodiversity, their traditional farming practices and their right to benefit from biological resources and related knowledge threatened by market forces. In particular, the World Commission on Environment and Development, among others, has stressed the importance of traditional knowledge in the sustainable development process and has observed that 'tribal and indigenous people will need special attention as the forces of economic development disrupt their traditional lifestyles' (World Commission on Environment and Development 1987, p. 12).

Accordingly, I turn to an examination of the localized commons in agriculture, which like the globalized commons in PGRFA is also under severe strain from international treaty regimes, which either articulate the rights and interests of governments and private economic operators, or which seek to accommodate the gains to be made from an increase in social welfare with the private rights and interests of individual farmers and farming communities.

3.1 The Changing Character of the Localized Commons in Agriculture

The localized commons in agriculture is best understood as one in which agricultural production for human consumption and the culture of food overlap (Echols 2002, pp. 17–22). The private sphere of agricultural production has traditionally been based on personal experience, trial and error, with traditions stretching back generations that are involved with the improvement of seed and livestock and often with very little or no public regulation, except when agricultural food or fibre was brought to market.

This stands in strong contrast to present-day farming in all industrialized countries and now increasingly developing countries, where farming has become a business as much as a way of life and is increasingly dominated by market monocultures (Echols 2002, pp. 37–40 and Shiva 1993, 39–40). In many developing countries, alongside traditional smallholders and local gardens, there has been an increase in the mechanization of agriculture and a re-orientation towards large-scale production of cash crops and heads of cattle for export markets. This is a process of gradual enclosure that has been hastened by fundamental changes in the way in which agricultural trade is being conducted globally and gradually being integrated into the multilateral trading system. In other words, the global market is encroaching upon the localized agricultural commons.

At first sight this particular act of enclosure seems strange not least because there is a commonly held belief that agricultural trade is regulated primarily by WTO rules and that the matter has been adequately treated to the benefit of all participating countries. However, until recently the regulation of agriculture in the multilateral trading system has been weak and ineffective mostly because for nearly half a century 'the GATT legal system [was not . . .] able to engage agricultural trade policy in a significant way' (Hudec 1993) and thus the opportunity was missed to bring about tariff reductions on agricultural products in comparison with industrial products. Instead governments have managed for decades to protect their domestic agricultural systems from penetration by market operators through various domestic support measures and the reasons for this are as follows.

Following a 1951 breach of GATT rules over dairy quotas (GATT/CP.6/SR.10, 1952), the United States was granted a waiver from its obligations under the General Agreement with respect to various aspects of its domestic agricultural programme under the Agricultural Adjustment Act of 1933, Decision of 5 March 1955. This had serious consequences for the liberalization of international agricultural trade among other GATT contracting parties. So too did the formation of the Common Agricultural Policy (CAP) by the European Economic Community (Snyder 1985), which sought to balance community preferences and rural development with an expansionist approach towards trade in agricultural products with its overseas territories and third country exporters (McMahon 2001, pp. 32–4).

These two significant events led to a virtual disengagement from the liberalization of agricultural trade by the contracting parties to the GATT, which has only recently been redressed as a result of the Uruguay Round Multilateral Trade Negotiations or MTN[27] and the adoption of the Agreement on Agriculture (Agreement on Agriculture 2001) at Marrakesh in 1994. A more market-orientated approach to the production and distribution of agricultural products globally has been endorsed by two other multilateral agreements that were concluded during the Uruguay Round MTN. The first is the Agreement on Sanitary and Phytosanitary Measures (SPS Agreement or SPS 2001) and the other is the Agreement on Trade-related Aspects of Intellectual Property Rights (TRIPs Agreement 2001). Each of these multilateral trade instruments can be read as a form of enclosure of the localized commons of agriculture, by means of encroachment on traditional agricultural practices and the rural livelihoods of many individual farmers and farming communities worldwide.

3.2 Enclosing the Localized Commons in Agriculture

One of the multilateral trade instruments that has had an enclosing effect on the localized commons is the Agreement on Agriculture, which is aimed at substantial progressive reductions in agricultural support and protection by Members, with the intention of bringing about a fundamental reform of international agricultural trade so as to correct and prevent restrictions and distortions in world agricultural markets (TRIPs Agreement 2001, Second Recital and Part XII, Article 20, pp. 39 and 55 respectively). In line with the Agreement on Agriculture Members must progressively liberalize their agricultural trade and expand market access and restrict the use of export subsidies and domestic support for agriculture.[28] Put differently, 'the exigencies of agricultural production, the social role of farming and the goal of food security no longer justify import restrictions and huge export subsidies' (Echols 2002, p. 41).

While this process can be read as one of market enclosure, Article 20(c) of the Agreement on Agriculture requires the Members to take into account the contribution that farming makes to a series of societal goals, or non-trade concerns. As the matter currently stands, there is basic agreement among governments that every country has the right, in accordance with mutually agreed rules, to address non-trade concerns, such as strengthening the socioeconomic viability and development of rural areas, food security and environmental protection, and promoting the coexistence of various types of agriculture (Note on Non-trade Concerns 2000). However, there is considerable difference among them as to how to achieve this and in such a way that countries do not use Article 20(c) in order to disguise domestic support for agricultural production, by shifting the focus of the reform debate away from 'pure trade' concerns to broader benefits that society may derive from agriculture.

Some Members embrace the broader notion of 'multifunctionality' in agricultural trade reform (Anderson 2000 and Smith 2000), which has its origins in the FAO concept of sustainable agriculture and rural development or SARD (FAO Glossary and Hardaker 1997) and multi-functional agriculture is the sort of agriculture that 'conserves land, water, plant and animal genetic resources, in environmentally non-degrading, technically appropriate, economically viable and socially acceptable' ways (Faber 1999).

The European Communities and Japan are champions of multifunctionality, as are Norway and South Korea. Both the European Commission and the Japanese Government believe that it is in the interest of society and the individual for them to manage markets in such a way that they generate desirable and practicable public goods. For example, the European Communities as the world's biggest importer and second largest exporter of agricultural products considers the agricultural negotiations as an opportunity to further the European model of agriculture.[29] Encouraged by the failure of the Seattle Ministerial Meeting in 1999 to deliver a new round of trade negotiations and to demonstrate broad public concern about the potential impact of unbridled globalization on the environment, health standards and cultural diversity, the European Communities is of the view that a 'multifunctional, sustainable and competitive farming system' can address these concerns (EU Agriculture and the WTO 2001).

It argues that some forms of support are needed in order to maintain agricultural activity in relation to environmental and rural functions in rural areas. In the view of the European Communities, agriculture can continue to play an important role in the protection of the environment, by protecting biodiversity, including the diversity of ecosystems and habitats, species diversity and genetic diversity. Moreover, it can contribute to safeguarding the scenic features of traditional cultivated landscapes, which may have aesthetic, cultural and historical value that contribute to the well-being of society (European Commission, DG Agriculture 1999, p. 2).

This latter aspect has also been recognized in the European Landscape Convention (2000), adopted by the Members of the Council of Europe, which is devoted to the protection, management and planning of *all* landscapes in Europe. It explicitly recognizes in its preamble that 'the landscape has an important public interest role in the cultural, ecological, environmental and social fields' besides '[constituting] a resource favourable to economic activity' (European Landscape Convention 2000, Third Recital). The Convention goes on to describe the contribution that 'the landscape contributes to the formation of local cultures' and that it is 'a basic component of the European natural and cultural heritage', which is 'an important part of the quality of life for people everywhere' (European Landscape Convention 2000, Fourth Recital).

The European Communities approach on multifunctionality also considers government intervention to be necessary in order to reward the production of services related to food safety and quality, which is of prime concern to consumers. While such services benefit the public at large they are not sufficiently rewarded by the market, nor are they reflected in the price of products (European Commission, DG Agriculture 1999, pp. 2–3).

Other WTO Members, particularly those comprising the Cairns Group[30] of non-subsidizing, agricultural exporting countries do not support the concept of multifunctionality but instead take a more restrictive view towards non-trade concerns. They want to see all non-tariff barriers, other than quarantine regulations, transformed into tariffs and bound into GATT schedules and to have farm production and export subsidies reduced and eventually phased out (Anderson 2001, p. 89). Since they have largely committed themselves domestically to dismantling domestic support measures, they see the debate on multifunctionality as going nowhere. From an ideological and practical standpoint, they consider it to be driven by national self-interest and to believe that the market will in time deliver.

While not rejecting the need to address non-trade concerns outright in future negotiations, the Cairns Group Members consider for example that environmental benefits, like any other social benefits, will flow only once there is complete liberalization of trade through economic growth and enhanced welfare, and that environmental payments to farmers could substantially distort the market if paid on a large scale. Should there be any need to deal with non-trade concerns then they must be considered under the 'green box' measures, currently considered as the least trade distorting under the Agreement on Agriculture (Sensi and Werksman 2001, pp. 474–6). Hence, this localized agricultural commons is not only under threat from the encroachment of global trade rules on the localized agricultural commons but also this process is subject to a set of competing claims over non-trade concerns.

Another form of enclosure is discernible from the way in which the phenomenon of global market for agricultural production ensures that agriculture is linked more closely to science, technology and economics than local culture (Sensi and Werksman 2001, pp. 474–6). This has the result that government intervention has increased markedly, including government policy to control the supply of food and to ensure its safety (quarantine, testing requirements, controlled use of pesticides and additives and so on) by means of a process of enclosure, which is market-driven. The process is reinforced by the SPS Agreement (SPS Agreement 2001), which permits WTO Member governments to apply certain sanitary and phytosanitary standards for the protection of human, animal and plant life or health provided this is not done in a discriminatory or arbitrary manner and the measures do not constitute a disguised restriction on trade (SPS Agreement 2001, Article 2).

On the face of it the SPS Agreement may seem innocuous but the encroachment of market forces on the domestic regulatory sphere has had the effect that Members are being encouraged to harmonize their domestic sanitary and phytosanitary standards upwards in an effort to comply with relevant international standards (Echols 2002, p. 42). Thus, in some respects the boundaries for domestic policy-making and subsequent regulation are being set in international decision-making fora such as the WTO and the related FAO/WHO Codex Alimentarius[31] that has the attendant effect of enclosing the localized commons in agriculture.

It has even been argued that the SPS Agreement 'uses science to shift the balance from the GATT's loosely fettered sovereignty to regulated sovereignty [under the WTO]' (Echols 2002, p. 42). This is because governments are obliged to rely on science when developing a sanitary or phytosanitary measure, which must be 'based on scientific principles' (SPS Agreement 2001, Article 2:2), shall not be maintained without 'sufficient scientific evidence' (SPS Agreement 2001, Article 2:2) and shall be based on a 'risk assessment' (SPS Agreement 2001, Article 5:1; World Trade Organization 1998).

Another feature of the enclosure of the localized agricultural commons is the shift to genetically modified or GM crops. In the mid-1990s virtually no genetically modified food was produced and now a decade later, we are witnessing unlimited use of GM crops in North America, particularly in such staple crops as wheat and corn in the United States, with an expansion of that process in major developing country exporters, such as Argentina,[32] China, India, the Philippines and South Africa (Larson 2003). Until we can be certain of the viability of GM innovations (Brac de la Perrière and Rao 2000), this process of enclosure remains perilous. But perhaps what is more disconcerting is that ultimately the turn towards GM crops is yet another sign of the homogenization, or standardization of some essential agricultural crops that potentially may lead to a further loss of agrobiodiversity (Footer 2004, p. 434).

A further act of enclosure, which is accompanying the agricultural trade reform process and is operating alongside the public regulation of the market, is the encroachment of proprietary rights of all kinds, but particularly intellectual property rights, on farming techniques and agricultural practices. The situation of local farmers in developing countries is particularly critical in this respect.

Traditionally, local and indigenous communities in developing countries have bred and developed their crop varieties, improved on those varieties through selective breeding and sold them locally under names that have found widespread local acceptance but in recent times the encroachment of market forces has encouraged many farmers to export local varieties. What is often not appreciated is that despite or because of international treaty instruments like the TRIPs Agreement, which provides a common set of rules to protect and enforce intellectual property rights among WTO Members, intellectual property protection in developing countries remains weak; some governments are even opposed to the grant of monopoly rights over agricultural crops. This can lead to an agricultural biotechnology company from a developed country acquiring samples of crop germplasm and, where these varieties are not produced using biotechnology, utilizing the exemption for plants and animals, which is provided in Article 27:3(b) of the TRIPs Agreement (TRIPs Agreement 2001, Article 27:3(b)), and claiming that these are natural varieties where no inventor can be identified (Footer and Opuku Awuku 2005).

Since many developing countries do not provide intellectual property protection for their plant varieties – not even protection under one of the Acts of the UPOV (International Union for the Protection of New Varieties of Plants 1972, either the Act of 1978 or the Act of 1991; Footer and Opuku Awuku 2005) – an agricultural biotechnology company can genetically engineer a close substitute for the natural variety, maintaining its desirable consumer characteristics. The next step is to patent the genetically modified variety and to copyright its names, thereby making it eligible for intellectual property protection under the TRIPs Agreement. This procedure has been adopted by private biotechnology firms that are intent on licensing the production of crops in climatically friendly countries, exporting the product in competition with natural varieties and preventing the natural varieties from being sold in importers' markets using their natural names (Kerr et al. 1999). The best-known example of this practice to date are Jasmine rice from Thailand and Basmati rice from India, both of which are rice varieties that have been patented and copyrighted by United States firms (Kerr et al. 1999).

From one view this practice – sometimes labelled 'biopiracy' (Shiva 1999) – cannot be the basis for the grant of patents or the establishment of exclusive marketing rights because a patent system that is intended to reward inventiveness and creativity systematically but fails to apply the criteria of novelty and non-obviousness honestly, especially in relation to traditional knowledge and innovations of local and indigenous farming communities, is flawed and should be changed. On another view the extension of intellectual property rights protection to living forms, including plants, plant varieties and their genetic information is a valid exercise and is supported by international treaty instruments.[33]

While it is true that Article 27:3(b) of the TRIPs Agreement allows for the protection of plant varieties through an 'effective *sui generis* system', and could therefore take account of the rights of rural communities, this may not prove adequate in the long run for many countries since it is very unclear what such a system should entail, nor does it necessarily offer enough safeguards to local and indigenous farmers for the protection of their traditional varieties and the practices associated with them, which may include knowledge related to traditional methods of agricultural production (Footer and Opuku Awuku 2005). In effect the process that I have just described currently allows a WTO Member 'to prohibit the patenting of a "non-modified" plant but it absolutely must protect the intellectual property rights of an inventor who has crafted a plant variety, by, for instance, inserting a foreign gene into a plant' (Brac de la Perrière and Rao 2000, p. 94).

Finally, these various acts of enclosure are facilitated by the process of globalization, by which I understand the ways in which global human interaction, interconnectedness and awareness among peoples is intensified. Globalization 'signals the growing interdependence and interpenetration of human relations alongside the increasing integration of the world's socioeconomic life' (Webster 1995, p. 141 and Habermas 1998, pp. 101–5). The debate on non-trade concerns manifests this aspect of globalization but at the same time presents a different form of enclosure whereby the localized commons in agriculture is potentially bounded and contained by public regulation.

In this respect, enclosure of the localized commons in agriculture differs from the enclosure of the global commons of PGRFA because it is an encroachment of public rights and duties on the private sphere rather than private rights on the public sphere. However, what both sets of commons share is that the process of enclosure in both is largely a regulatory one, which is made operational through a set of international treaty instruments and policy and institutions that have been developed within the framework of those instruments.

4 CONCLUSIONS

Having noted some of the similarities and differences that exist between the global commons for PGRFA and the localized agricultural commons, the question arises as to whether it is possible to retain, or rebuild, a global commons for the benefit of society at large, or even humankind, and if so, what form should this take? Of course none of this is particularly new. After all, the enclosure of the commons of nature has continued relentlessly for centuries but perhaps we should stop for a moment to consider our obsession with ownership and all forms of commodification.

The global commons in PGRFA is non-replenishable or may be irreparably harmed by excessive use (an example of the 'tragedy of the commons' [Hardin 1968], whereby users of the commons are caught up in the process of destroying the very resources on which they depend). The tragedy of the commons in this instance can also be read as the 'tragedy of open access' (Dutfield 1999) or the 'tragedy of enclosure' (Monbiot [1994] 1999).

Other commons like the localized commons in agriculture, if non-trade concerns are not fully appreciated and factored into agricultural trade reform, will fail to sustain and preserve rural landscapes and our agricultural heritage. Multilateral trade instruments in the fields of agriculture, public health and intellectual property protection form the new boundaries for a form of enclosure that is driven by market forces and impinges deeply on traditional domestic regulatory spheres of competence in the maintenance of farming techniques and practices, rural landscapes and sustainable development.

NOTES

* This chapter is a revised and updated version of a presentation that the author gave at the Australian Centre for Intellectual Property and Agriculture (ACIPA), Australian National University (ANU), Canberra, on 22 August 2002.

1. PGRFA is the genetic material of plants used for agricultural purposes that is of value as a resource for present and future generations of people. As a sub-species of the wider concept of genetic resources (and plant genetic resources or PGR) they form part of what is commonly known as biological diversity (or 'biodiversity'). Thus, PGRFA form part of (1) an ecosystem, (2) are a sub-species (of genetic resources) and (3) constitute 'diversity within species, between the species and of ecosystems', in the sense of Article 2 of the Convention on Biological Diversity 1992 (hereinafter CBD).
2. Footer (2000), where the term 'private ownership of the genetic commons' is borrowed from the concept as used by Frow (1997), Chapter 3.

3. From an economic perspective public goods have been defined as 'collective consumption goods . . . which all enjoy in common in the sense that each individual's consumption of such a good leads to no subtraction from any other individual's consumption of that good'. Public goods are characterized as non-rivalrous and non-excludable; see Samuelson (1954).
4. These include not only patents but also other intellectual property rights such as plant breeders' rights (or PBRs) and various forms of restrictive licensing.
5. Often referred to as 'the enclosure movement', the process involved a series of enclosures of common land that continued with differing means, ends and varieties of state involvement between the fifteenth and nineteenth centuries. See Yelling (1977), cited in Boyle (2003), 33–74.
6. Many of the observations about commodification and its relationship to PGRFA draw on the second part of the essay by Frow (1996), 94; Frow (1997), p. 198, who in turn relies upon the observations of Kloppenburg (1988), pp. 9–11. See also Shiva et al. (1997).
7. Many of the world's staple crops, which now dominate the agricultural economies of industrialized nations, are not indigenous to North America or Europe but are found in the so-called Vavilovian centres of genetic diversity (named after the Russian botanist N.I. Vavilov who first charted the centres of origin of the world's most important crops, almost all of which are in the developing world). They include wheat and barley developed by the peoples of Asia Minor, the Mediterranean and Ethiopia; corn and potato cultivated by the Inca and other Indian tribes in the high Andes; and rice and soybean developed in China.
8. The CGIAR consists of an informal association of 57 public and private sector members, drawn from developed and developing countries, private foundations and development banks and is based upon an informal arrangement between the World Bank, the UN Development Programme (UNDP), and the FAO; see Footer (2000), pp. 56–7.
9. Ex situ conservation is 'the conservation of components of biological diversity outside their natural habitats', CBD (1992). It is distinguishable from in situ conservation, which in that same provision is described as 'the conservation of ecosystems and natural habitats and the maintenance and recovery of viable populations of species in their natural surroundings and, in the case of domesticated or cultivated species, in the surroundings where they have developed their distinctive properties', CBD (1992).
10. Examples of ex situ genebanks are diverse but they may also include botanical gardens' collections; see for an overview Johnston (1993).
11. Pardo and Christol (1986) have identified five basic elements that form the content of the common heritage of humankind principle: (1) natural resources (or their spatial location) cannot be subject to national appropriation; (2) all states should share in the management of these scarce resources; (3) there should be actual sharing of benefits derived from any resource exploitation; (4) the dedication of the natural resources to peaceful purposes; and (5) the preservation of the space for future generations. See Pardo and Christol (1986).
12. On spontaneous and directed orders, see Roessler (1978), citing Von Hayek (1973), pp. 36–8.
13. See Ten Kate and Laird (1999), p. 119, who note the extensive network of bilateral aid projects designed to introduce elite germplasm from developed into developing countries, for example, the introduction of soybean germplasm from USDA into Brazil, which formed the basis for the Brazilian soybean economy and high lysine maize.
14. This process was gradual and not free of controversy (UN General Assembly 1962; see Brownlie 1983, pp. 176–9). Then in the early 1970s eventually two UN General Assembly Resolutions were adopted by consensus on 1 May 1974 (UN General Assembly Resolutions 1974).
15. See Footer (2000), pp. 58–61, for extensive discussion of this process.
16. FAO Resolution 8/83, 1983; see Footer (1999) pp. 62–8 for a preliminary overview.

17. For a brief analysis of management regimes that embody the common heritage principle (now more usually the 'common concern of humankind' principle), see Franck (1995), pp. 75–9.
18. FAO Resolution 3/01, 2001, Preamble (Third Recital). This reflects the preambular text of the CBD (Convention on Biological Diversity, 5 June 1992).
19. FAO Resolution 3/01, 2001, Article 11, Paragraphs 1 and 2. The actual listing of the 35 crop genera and 29 forage species is contained in Annex I to the Treaty.
20. FAO Resolution 3/01, 2001, Article 12.3(a), which specifically states that this should not include 'chemical, pharmaceutical and/or other non-food/feed industrial uses'.
21. Countries like the US, Canada and Japan strongly opposed the inclusion of limitations on the grant of intellectual property rights, fearing that this would stifle innovation.
22. On the notion of 'corrective equity', see Franck (1995), pp. 75–9.
23. FAO Resolution 3/01, 2001, Article 14. The FAO Global Plan of Action for the Conservation and Sustainable Use of Plant Genetic Resources for Food and Agriculture was originally set up in 1983 as one of the instruments in the overall FAO Global System (FAO Resolution C 9/83, 1983), and was formally adopted by 150 countries, following the adoption of the FAO (FAO Report 1996). The main task of the Global Action Plan is to provide a periodic reporting system (by FAO Member governments) in order to keep track of the state of the world's plant genetic resources.
24. An Expert Group on the Terms of the Standard Material Transfer Agreement, or MTA, charged with developing the relevant contractual basis for access and transfer of individual crop germplasm in the Multilateral System, has submitted its Report of the Expert Group 2004.
25. FAO Resolution 3/01, 2001, Article 13, Paragraph 2(d)(ii), first full paragraph, in conjunction with Article 19, Paragraph 3(f).
26. FAO Resolution 3/01, 2001, Article 15, Paragraphs 1(a) and 2. Crop germplasm in such ex situ collections, which does *not* appear on the Annex I list, for example soya, and which was collected *before* the date of entry into force of the Seed Treaty, will continue to be administered through a set of FAO/CGIAR agreements (and the relevant MTA) in order to facilitate access. It is intended that the Governing Body will eventually monitor future use of the particular MTA under those agreements for a period of four years from 29 June 2004 onwards.
27. The Uruguay Round MTN was launched at the Ministerial Meeting held in Punta del Este, Uruguay; see Declaration of Punta del Este (1987).
28. See McMahon (2001), Chapter 1 *in extenso* for the historical background that informs the agricultural reform process.
29. The centrepiece of this European model of agriculture is of course the Common Agricultural Policy or CAP, major revisions to which are slowly being put into place; see for details www.europa.int.
30. Named after the Australian city where the first meeting of the group was held, the Cairns Group currently consists of Australia, Argentina, Brazil, Canada, Chile, Colombia, Fiji, Indonesia, Malaysia, New Zealand, Paraguay, Philippines, South Africa, Thailand and Uruguay.
31. See for example, Organisation for Economic Co-operation and Development (OECD) (1999), where it is clear that domestic policy-making is increasingly influenced by decision-making in international fora, on the basis of treaty obligations that are having a profound impact on domestic regulatory systems in such areas as sanitary and phytosanitary measures (SPS).
32. Argentina, alongside the United States and Canada, is one of the Parties that have brought the current dispute (World Trade Organization 2004) – the Panel Report was issued on 29th September 2006.
33. See the Landmark US Supreme Court Decision 1980, where it was determined that patents could be granted for living organisms and this led to the rapid extension of patents to complex organisms including plants and animals. See also European Parliament and Council Directive 98/44/EC 1998.

REFERENCES

Agreement on Agriculture (2001), *The Legal Texts*, pp. 39–68.

Anderson, K. (2000), 'Agriculture's "Multifunctionality" and the WTO', *Australian Journal of Agricultural and Resource Economics*, **44**(3), September, 475–94.

Anderson, Kym (2001), 'Developing Country Interests in Agricultural Trade Reform: A Cairns Group Perspective', in Joseph A. McMahon (ed.), *Trade & Agriculture: Negotiating a New Agreement*?, London: Cameron, pp. 89–120.

Boyle, J. (2003), 'The Second Enclosure Movement and the Construction of the Public Domain', *Law and Contemporary Problems*, **66**(1&2), 33–74.

Brac de la Perrière, Robert Ali and Vijaya Rao (2000), *Brave New Seeds: The Threat of GM Crops to Farmers*, London and New York: Zed Books.

Brownlie, Ian (ed.) (1983), *Basic Documents in International Law*, 5th edition, Oxford: Oxford University Press.

Convention on Biological Diversity (CBD) (1992), Article 2, 5 June 1992, *International Legal Materials*, **31**(818), in force 29 December 1993.

Cullet, Ph. (2001), 'Property Rights over Biological Resources', *Environmental Planning C: Government and Policy*, **19**, Cambridge, MA: Harvard University Press, 651 et seq., www.ielrc.org/content/a0104.pdf.

Declaration of Punta del Este (1987), Ministerial Meeting, GATT, *Basic Instruments and Selected Documents* (BISD), 33S/19.

Declaration of the United Nations Conference on the Human Environment (1972), 'Report of the United Nations Conference on the Human Environment', UN Doc. A/CONF.48/14/Rev.1, Stockholm, 5 June 1972, *International Legal Materials*, **11**(1416), 39–45.

Dutfield, G. (1999), 'Sharing the Benefits of Biodiversity: Access Regimes and Intellectual Property Rights' (*Science, Technology and Development Discussion Paper*), Cambridge, MA: Center for International Development and Belfer Center for Science and International Affairs, Harvard University.

Echols, Marsha A. (2002), *Food Safety and the WTO: The Interplay of Culture, Science and Technology*, Boston, The Hague, New York: Kluwer Law International.

EU Agriculture and the WTO (2001), www.europa.eu.int/comm/dg06/index, cited in Sensi and Werksman 2001, p. 470.

European Commission, DG Agriculture (1999), 'Safeguarding the Multifunctional Role of EU Agriculture: Which Instruments'; www.europa.eu.int/comm/dg06/index.

European Landscape Convention (2000), *European Treaty Series*, **176**, Florence, 20 October 2000, in force 1 March 2004.

European Parliament and Council Directive 98/44/EC (1998), On the Legal Protection of Biotechnological Inventions, *Official Journal of the European Communities*, L 213/13, 6 July 1998.

Faber, Gerrit (1999), 'Cultivating Our Futures. Issues Paper: The Multifunctional Character of Agriculture and Land', Paper prepared for the FAO/Netherlands Conference on The Multifunctional Character of Agriculture and Land, Maastricht, 12–17 September 1999, available at http://www.fao.org/waicent/search/simple_s_result.asp.

FAO Glossary, 'FAO Glossary and Acronyms: Trade in Agriculture, Fisheries and Forestry'; http://www.fao.org/trade/glos_en.asp#s.

FAO Report (1996), *Report on the State of the World's Plant Genetic Resources for Food and Agriculture*, prepared for the Fourth International Technical Conference on Plant Genetic Resources, Leipzig, Germany, 17–23 June 1996, and known as the 'Leipzig Declaration'.

FAO Resolution 8/83 (1983), The International Undertaking on Plant Genetic Resources (International Undertaking, or IU), adopted by FAO Conference, 22nd Session, Rome, 5–23 November 1983.

FAO Resolution C 9/83 (1983), FAO Global System for the Conservation and Utilisation of Plant Genetic Resources for Food and Agriculture, adopted by FAO Conference, 22nd Session, Rome, 5–23 November 1983.

FAO Resolution C 4/89 (1989), Agreed Interpretation of the International Undertaking, adopted by FAO Conference, 25th Session, Rome, 11–29 November 1989, Paragraph 2; ftp://ext-ftp.fao.org/waicent/pub/cgrfa8/Res/C4-89E.pdf.

FAO Resolution 5/89 (1989), Agreed Interpretation of the International Undertaking, Farmers Rights, adopted by FAO Conference, 25th Session, Rome, 11–29 November 1989.

FAO Resolution 3/91 (1991), adopted by FAO Conference, 26th Session, Rome, 9–27 November 1991; available at: ftp://ext-ftp.fao.org/waicent/pub/cgrfa8/Res/C3-91E.pdf.

FAO Resolution C 7/93 (1993), adopted by FAO Conference, 27th Session, Rome, 6–24 November 1993.

FAO Resolution 3/01 (2001), 'International Treaty on Plant Genetic Resources for Food and Agriculture' or 'Seed Treaty', adopted on 3 November 2001, FAO Conference, 31st Session, Rome, 2–13 November 2001, in force 29 June 2004; www.fao.org/waicent/faoinfo/agricult/cgrfa/IU.html.

Food Ethics Council (2002), 'TRIPS with Everything? Intellectual Property and the Farming World', 5th Report, Southwell, Notts.

Footer, M.E. (2000), 'Intellectual Property and Agrobiodiversity: Towards Private Ownership of the Genetic Commons', in Jutta Brunnée and Ellen Hey, *Yearbook of International Environmental Law*, **10**, Oxford: Oxford University Press, pp. 48–81.

Footer, Mary E. (2004), 'Our Agricultural Heritage: Sustainability, Common Heritage and Intergenerational Equity', in Nico Schrijver and Friedl Weiss (eds), *International Law on Sustainable Development: Principles and Practice*, The Hague, London: Martinus Nijhoff, pp. 433–66.

Footer, Mary E. and Emmanuel Opuku Awuku (2005), 'Sustainable Agricultural Resources and Food Security: The Seed Treaty and Equitable Benefit Sharing', in Marie-Claire Cordonier Segger and Judge C.G. Weeramantry (eds), *Sustainable Justice: Integrating Economic, Social and Environmental Law*, Leiden, Boston: Martinus Nijhoff, pp. 241–56.

Franck, Thomas M. (1995), *Fairness in International Law and Institutions*, Oxford: Clarendon Press.

Frow, J. (1996), 'Information as Gift and Commodity', *New Left Review*, **219** (September/October 1996), 89.

Frow, John (1997), *Time and Commodity Culture: Essays in Cultural Theory and Postmodernity*, Oxford: Clarendon Press.

GATT/CP.6/SR.10 (1952), 'US – Import Restrictions on Dairy Products', complaint brought by the Netherlands and Denmark, 24 September 1951, *Basic Instruments and Selected Documents* (BISD), 2nd Supplement (2S)/15, in Robert E. Hudec (1975), *The GATT Legal System and World Trade Diplomacy*, New York: Praeger Publishers, pp. 165–84.

Girsberger, Martin A. (1999), *Biodiversity and the Concept of Farmers' Rights in International Law: Factual Background and Legal Analysis*, Bern, Berlin, Bruxelles, Frankfurt am Main, New York, Wien: Peter Lang AG.

Habermas, Jürgen (1998), *Die Postnationale Konstellation*, Frankfurt am Main: Suhrkamp.

Hardaker, J. Brian (1997), *Guidelines for the Integration of Sustainable Agriculture and Rural Development into Agricultural Policies*, FAO Agricultural Policy and Economic Development Series 4, Rome: Food and Agricultural Organization of the United Nations.

Hardin, G. (1968), 'The Tragedy of the Commons', *Science*, **162**, 1243–8.

Hudec, Robert E. (1993), *Enforcing International Trade Law*, London: Butterworths, p. 336, cited in Joseph A. McMahon (ed.) (2001), *Trade & Agriculture: Negotiating a New Agreement*?, London: Cameron May, pp. 249–76.

International Union for the Protection of New Varieties of Plants (1972), 2 December 1961, *United Nations Treaty Series*, **815**(11609), in force 10 August 1968, subsequently amended by Acts of 10 November 1972, 23 October 1978, in force 8 November 1981 and 19 March 1991, in force 24 April 1998.

Johnston, S. (1993), 'Conservation Role of Botanic Gardens and Gene Banks', *Review of European Community & International Environmental Law* (RECIEL), **2**(172).

Kerr, W.A., J.E. Hobbs and R. Yampoin (1999), 'Intellectual Property Protection, Biotechnology and Developing Countries: Will the TRIPS be Effective?', *AgBioForum*, **2**(3 and 4), 203–11.

Kloppenburg, Jack Ralph Jr. (1988), *First the Seed: The Political Economy of Plant Biotechnology, 1492–2000*, Cambridge: Cambridge University Press.

Koo, Bonwoo, Philip G. Pardey and Brian D. Wright (2003), 'Conserving Genetic Resources for Agriculture: Counting the Cost', Brief 6 of the International Food Policy Research Institute (IFPRI) Series *Research at a Glance: Biotechnology and Genetic Resource Policies*, Washington, DC: University of Minnesota and IFPRI.

Landmark US Supreme Court Decision (1980), *Diamond, Commissioner of Patents and Trademarks* v. *Chakrabarty*, 447 US 303.

Larson, A. (2003), 'Trade and Development Dimension of U.S. International Biotechnology Policy', *An Electronic Journal of the U.S. Department of State*, **8**(3), September 2003; http://usinfo.state.gov/journals/ites/0903/ijee/larson.htm.

McMahon, Joseph A. (ed.) (2001), *Trade & Agriculture: Negotiating a New Agreement*?, London: Cameron May.

Mgbeoji, I. (2003), 'Beyond Rhetoric: State Sovereignty, Common Concern, and the Inapplicability of the Common Heritage Concept to Plant Genetic Resources', *Leiden Journal of International Law*, **16**, 821–37.

Monbiot, George [1994] (1999), 'The Tragedy of Enclosure', in Darrell Addison Posey (ed.), *Cultural and Spiritual Values of Biodiversity*, London: UNEP and Intermediate Technology Publications.

Nairobi Conference (1992), Resolution 3 of the Nairobi Conference for the Adoption of the Agreed Text of the Convention on Biological Diversity, and relating to 'The Interrelationship between the Convention on Biological Diversity and the Promotion of Sustainable Agriculture' (22 May 1992), *International Legal Materials*, **31**(846).

Note on Non-trade Concerns (2000), Revision, Submission to the Special Session of the WTO Committee on Agriculture by Barbados, Burundi, Cyprus, Czech

Republic, Dominica, Estonia, the European Communities, Fiji, Iceland, Israel, Japan, Korea, Latvia, Liechtenstein, Madagascar, Malta, Mauritania, Mongolia, Norway, Poland, Romania, Saint Lucia, Slovak Republic, Slovenia, Switzerland and Trinidad and Tobago, G/AG/NG/W/36/Rev.1, 9 November 2000.

Ntambirweki, J. (2001), 'Biotechnology and International Law within the North–South Context', *The Transnational Lawyer*, **14**, 103–28.

Organisation for Economic Co-operation and Development (OECD) (1999), The Summary of 'Informal DAC Experts Meeting on Capacity Development for Trade', held on 23–24 November 1998, Document DCD/DAC(99) 13, at 5, Paragraph 12; www.oecd.org/dac/.

Pardo, A. and C.Q. Christol (1986), 'The Common Interest: Tension Between the Whole and the Parts', in Ronald Saint John McDonald and Douglas M. Johnston (eds), *The Structure and Process of International Law*, Dordrecht: Kluwer Law, p. 654.

Raustiala, K. and D.G. Victor (2004), 'The Regime Complex for Plant Genetic Resources', *International Organization*, **58** (Spring, 2004), 277–309.

Redgwell, Catherine (1999), *Intergenerational Trusts and Environmental Protection*, Manchester: Manchester University Press, 129–132.

Report of the Expert Group (2004), 'Report on the Outcome of the Expert Group on the Standard Terms for the Material Transfer Agreement' to the Second Meeting of the CGFRA, acting as the Interim Committee for the International Treaty on Plant Genetic Resources for Food and Agriculture, Rome, 15–19 November 2004, CGRFA/IC/MTA-1/04/Rep.; www.fao.org/ag/cgrfa/docsic2.htm.

Robinson, N.A. (ed.) (1993), *Agenda 21: Earth's Action Plan*, New York: Oceana Publications, www.un.org/esa/sustdev/agenda21text.htm.

Roessler, F. (1987), 'Law, De Facto Agreements and Declarations of Principle in International Economic Relations', *German Yearbook of International Law*, **21**(27).

Safrin, S. (2004), 'Hyperownership in a Time of Biotechnological Promise: The International Conflict to Control the Building Blocks of Life', *American Journal of International Law*, **98**, 641–85.

Samuelson, P.A. (1954), 'The Pure Theory of Public Expenditure', *Review of Economics and Statistics*, **11**, 387–9.

Sensi, Stefano and Jacob Werksman (2001), 'Inside the Green Box: "Multi-functional" Agriculture and the Protection of the Environment', in Joseph A. McMahon (ed.) (2001), *Trade & Agriculture: Negotiating a New Agreement?*, London: Cameron May, pp. 463–83.

Shiva, Vandana (1993), *Monocultures of the Mind*, London and New York: Zed Books/Penang, Malaysia: Third World Network.

Shiva, V. (1999), 'Biopiracy: Need to Change Western IPR Systems', *The Hindu*, 28 July 1999, New Delhi, http://www.hinduonline.com.

Shiva, Vandana, Afsar H. Jafri, Gitanjali Bedi and Radha Holla-Bhar (1997), *The Enclosure and Recovery of the Commons: Biodiversity, Indigenous Knowledge and Intellectual Property Rights*, New Delhi: Research Foundation for Science, Technology and Ecology.

Singh Nijar, Gurdial (1999), 'Sui Generis Law for Plant Varieties: Preserving the Knowledge and Creativity of Traditional Breeders – a Third World View', *Third World Network*, 22 April 1999, http://www.twnside.org.sg/access.htm.

Smith, F. (2000), 'Multifunctionality and Non-trade Concerns in the Agriculture Negotiations', *Journal of International Economic Law*, **3**(4), 707–13.

Snyder, Francis (1985), *Law of the Common Agricultural Policy*, London: Sweet & Maxwell.

SPS Agreement (2001), Agreement on the Application of Sanitary and Phytosanitary Measures, *The Legal Texts*, pp. 59–72.

Ten Kate, K. and C.L. Diaz (1997), 'The Undertaking Revisited: A Commentary on the Revision of the International Undertaking on Plant Genetic Resources for Food and Agriculture', *Review of European Community & International Environmental Law* (RECIEL), **6**(3), 284–92.

Ten Kate, Kerry and Sarah A. Laird (1999), *The Commercial Use of Biodiversity: Access to Genetic Resources and Benefit-sharing*, London: Earthscan Publications Ltd.

Ten Kate, K. and S.A. Laird (2000), 'Biodiversity and Business: Coming to Terms with the "Grand Bargain" ', *International Affairs*, **76**(1), 241–64.

Thornström, C.-G. (2005), 'Producing International Public Goods in a Proprietary Science World – the CGIAR Contribution', Comment, No. 1, *BRIDGES*, January, 21–2.

Thrupp, L.A. (2000), 'Linking Agricultural Biodiversity and Food Security: The Valuable Role of Agrobiodiversity for Sustainable Agriculture', *International Affairs*, **76**(2), 265–81.

TRIPs Agreement (2001), Agreement on Trade-related Aspects of Intellectual Property Rights, *The Legal Texts*, pp. 379–80.

UN General Assembly (1962), 'Permanent Sovereignty over Natural Resources', GA Resolution 1803 (XVII), 14 December 1962.

UN General Assembly Resolutions (1974), 'Declaration on the Establishment of a New International Economic Order', GA Resolution 3201 (S-VII), 1 May 1974, and 'Programme of Action on the Establishment of a New International Economic Order', GA Resolution 3202 (S-VII), 1 May 1974.

Von Hayek, Friedrich A. (1973), *Law, Legislation and Liberty: A New Statement of the Liberal Principles of Justice and Political Economy, Volume 1: Rules and Order*, London: Routledge and Kegan Paul.

Webster, Frank (1995), *Theories of the Information Society*, London and New York: Routledge.

World Commission on Environment and Development (1987), *Our Common Future*, Oxford: Oxford University Press.

World Trade Organization (1998), 'European Communities – Measures Concerning Meat and Meat Products', Report of the Appellate Body (WT/DS26/AB/R, WT/DS48/AB/R/), adopted 13 February 1998 (*EC Hormones*), Paragraphs 177 and 180.

World Trade Organization (2004), 'European Communities – Measures Affecting the Approval and Marketing of Biotech Products' (WT/DS291, WT/DS292 and WT/DS293).

Yelling, James Alfred (1977), *Common Field and Enclosure in England, 1450–1850*, London: Macmillan.

PART IV

Regulating biotechnology through the patent system

10. Should we regulate biotechnology through the patent system? The case of terminator technology

Graham Dutfield

1 INTRODUCTION

According to the Devil in George Bernard Shaw's *Man and Superman*, 'in the arts of life man invents nothing; but in the arts of death he outdoes Nature herself, and produces by chemistry and machinery all the slaughter of the plague, pestilence and famine' (Shaw 1903 [2000]). Such a dark vision of human nature seems to sum up the views of many people that oppose the patenting of life forms, genetic modification and industrial agriculture, except that the blame is more likely to be placed at the door of industrial capitalism rather than of shortcomings inherent to human beings. This becomes evident when the same people extol the virtues of indigenous peoples and others 'embodying traditional lifestyles', in the language of the Convention on Biological Diversity (CBD 1992), for having a more environmentally friendly lifestyle than the rest of us, and for giving so much to the wellspring of human knowledge without getting a cent in return. At the other extreme we have the 'techno-optimists' who have a much stronger faith in our inherent creativity to improve on what we have inherited from nature and from past generations of humans.

The so-called terminator technology, or technologies (since there is a growing number of them), would superficially appear to reinforce the Devil's point of view. Terminator technology, as its name suggests, was coined not by proponents but by a Canadian activist, Pat Mooney, from an organization then known as Rural Advancement Foundation International (RAFI),[1] who was seeking to direct negative publicity towards it. In this he was highly successful.

Terminator first came to Mooney's attention in 1998 when he saw an announcement that a patent had been granted jointly to the United States Department of Agriculture (USDA) and Delta and Pine Land, a major American cotton seed company, describing molecular biological techniques for controlling gene expression in plants, plant parts or seeds so that traits can be switched on and off between generations.[2] Conceivably, farmers could benefit from these techniques, depending upon the traits in question whose expression or non-expression may help determine the success of the harvest. But among the claims is a method for producing seed that is incapable of germination, or to be more specific, a technology that would render harvested seed sterile. On the face of it, it seems extraordinary to invest so much effort and expense in developing a means to produce sterile seed. But despite the involvement of a public sector institution, this is strictly business. The purpose is to prevent farmers from replanting saved seed and thereby undercut seed company monopolies. In doing so, it provides a means not only of preventing the infringement of intellectual property protection but of ensuring the continuation of the monopoly beyond the life of any patent or plant variety certificate, assuming such activities require the authorization of the right holder in question. Not only this, but terminator technology has grave implications for the activity of breeding, which requires unrestricted access to plant varieties to be used as sources of initial variation.

The development of this technology seems to reflect the increased determination of the private sector (in this case, and in common with hybrids as we will see below, with the assistance of a public agency) to eliminate the replanting of proprietary seeds, which is also reflected in the increasing use of licensing agreements stipulating that customer-farmers must not replant their patent-protected seeds. Such agreements would, of course, become unnecessary if this technology became widely used. Genetic use restriction technologies (GURTs), of which terminator is just one, are not new. Hybridization is the original GURT. In the early twentieth century, a United States public sector plant scientist called George Shull discovered the phenomenon of (what he called) 'heterosis' in the corn plants resulting from his cross-breeding of inbred pure lines. This phenomenon, commonly referred to as 'hybrid vigour', is manifested in heightened yields. But because they are hybrids, the offspring cannot breed true and the yield enhancements thus last only for a single generation. So while farmers stand to benefit from seeds providing this hybrid vigour, they need to buy seeds at the beginning of every planting season to enjoy equally productive future harvests. This necessity was and continues to be a boon for the seed companies. Indeed, 'of the US$15 billion market in commercial seed at present, hybrids account for approximately 40 per cent of sales, and most of the profit' (Sehgal 1996). But as long as the cost of seed purchases is exceeded by additional revenues obtained through the hybrid vigour, farmers will continue to use hybrids in place of their open pollinating counterparts.[3]

The hybrid route to the breeding of better seeds is generally assumed to be a very good thing for the development of the seed industry, and in the opinion of many, but not all people, also for farmers. In fact, several of the world's major twentieth-century seed companies first came to prominence through their successful breeding of hybrid corn varieties. These include Pioneer Hi-Bred, DeKalb, Pfister and Funk. But sceptics argue that the massive investments in the development of hybrid varieties that were made in the 1920s and 1930s could have been allocated to breeding based on more conventional techniques that would have achieved similar yield increases but without preventing farmers from being able to replant their harvested seeds. Berlan and Lewontin are particularly negative (as they are about terminator), arguing that hybridization is a kind of 'deterioration technique' that not only enables seed companies to eradicate on-farm saving and exchange but actually eliminates all opportunities to improve crops through selective breeding (Berlan and Lewontin 1998). Farmers may gain in the short term, but widespread adoption of hybrid varieties may not necessarily best favour their long-term interests. Moreover, it may be true that breeders sometimes take advantage of lack of competition in certain seed markets profitably to deploy hybrid technology *without* needing to produce any significant productivity increases.

In the early days hybrid productivity was not much greater than their conventionally bred counterparts (Bugos and Kevles 1992). However, from the middle of the century, increased private investment had considerably improved the yields of hybrid corn. Unfortunately for breeders, hybridization does not work so easily for some of the most widely cultivated crops like wheat and rice, and is consequently less commercially viable. This, of course, presents problems for breeders. Plants are self-reproducing. With no law to prevent it, there is nothing to stop farmers from replanting harvested grain as seed, or even multiplying seed for the purpose of selling it in competition with the breeder (assuming this would be more profitable for them than selling harvested produce).[4]

Terminator technology appears to provide the solution to the problem (see Table 10.1). But unlike this earlier biotechnology protection system and other GURTs under development that seek merely to control the expression of specific traits, terminator, which uses seed sterility as the basis of its use restriction, provides no productivity or agronomic benefits to the farmer who buys the seed. Worse still, it is actually a net loss, since it removes a freedom but offers no compensating gains as did hybrid corn and as the so-called trait-specific GURTs promise to provide.

This point gives rise to some important questions. The most basic one is that of whether the terminator controversy affects the future of agriculture or is much ado about nothing? Second, if patenting is about promoting inventive activity for the benefit of the public, is terminator the kind of invention we should be encouraging in this way? Third, if not, are there

Table 10.1 Appropriating plant breeding innovations: legal and technological solutions

	Crop Type		
	Self-pollinators	Cross-pollinators	Asexual reproducers
Examples	Wheat, rice, sorghum	Maize, millet, pulses	Fruit trees, potatoes
Key features	Breed true	Do not breed true	Can be rapidly reproduced
Obstacle to appropriation	Harvested seed can be replanted	Deleterious effects of inbreeding	Easy to copy
Legal solution	PVP, contracts/ licences	Trade secrecy, contracts/licences	PVP, contracts/ licences
Continuing obstacles to appropriation	• Farmers' privilege • Research exemption • Difficult to enforce rights		• Farmers' privilege • Research exemption • Difficult to enforce rights
Technological solution	Terminator technology	Hybrids	

sufficient legal grounds for preventing its legal protection through the patent system? Finally, if these grounds are lacking, ought countries to ban such technologies by expanding the applicability of the *ordre public* and morality exclusions available to World Trade Organization (WTO) Member States by virtue of Article 27.3(b) of the Agreement on Trade-related Aspects of Intellectual Property Rights (TRIPS), according to which WTO 'Members may exclude from patentability inventions, the prevention within their territory of the commercial exploitation of which is necessary to protect *ordre public* or morality, including to protect human, animal or plant life or health or to avoid serious prejudice to the environment'? The rest of this chapter seeks answers to these questions.

It puts forward three arguments. First, countries have the sovereign right to determine whether or not it is appropriate to extend a patent monopoly to inventions it deems as morally objectionable or contrary to the public interest. Nonetheless, one must keep in mind that the commercialization of the technology in question is a matter for national regulatory and competition authorities to decide upon and not the patent-granting office.

Second, countries need to adopt rational, well-conceived and consistent competition, agricultural development and seed regulation policies. The objective of these policies should be to prevent GURTs from encouraging excessive concentration in the seed production and distribution markets, and to ensure that farmers can choose the seed they wish to plant, GURT-protected, modern, traditional or otherwise, without undue interference. Finally, but perhaps most importantly, as business becomes better able to maximize returns from its agro-biotechnological research outputs through legal and technological means, it becomes ever more vital to support public sector research targeted not just at commercial agriculture but also at poor subsistence farmers in the developing world.

2 HIGH STAKES

When even mild criticism of the technology irritates the US government enough for it to become heavy-handed with intergovernmental organizations, it becomes clear that the stakes are high. Terminator does matter, although it is difficult to be certain about its long-term impacts. The Conference of the Parties to the CBD has for several years been concerned about GURTs and at its sixth meeting in 2002 adopted a decision that, among other things, invited the International Union for the Protection of New Varieties of Plants (UPOV) to examine 'the specific intellectual property implications of genetic use restriction technologies, particularly in respect of indigenous and local communities'. The UPOV Office's memorandum, submitted to the CBD secretariat in January 2003, was a fairly tame document that expressed some mild scepticism of the benefits of GURTs but did not condemn them outright, and was mostly concerned to uphold the integrity of the UPOV system of plant variety protection (UPOV 2003a). The United States Patent and Trademark Office was nonetheless alarmed about this sufficiently to write to UPOV Vice Secretary-General Rolf Jördens expressing its objection to the submission of the document and its disagreement with the views expressed concerning GURTs, and requesting that it be withdrawn.[5] UPOV complied and in April 2003 sent the CBD Secretariat a rather bland position paper, which curiously barely even mentioned GURTs (UPOV 2003b).

Perhaps the main indicator of terminator's importance, though, lies in the fact that the USDA has been developing GURTs with the private sector as part of the US government's wider and long-term effort to protect the intellectual property of its businesses in overseas markets including developing countries. Indeed, according to a spokesman from the USDA, the aim is for the technology to be 'widely licensed and made expeditiously available to many seed companies' in order 'to increase the value of proprietary seed owned by US seed companies and to open up markets in Second and Third World countries' (RAFI 1998). Dr Harry Collins of Delta and Pine Land, co-owner of the patent with the USDA, claimed that the patent 'has the prospect of opening significant worldwide seed markets to the sale of transgenic technology for crops in which seed currently is saved and used in subsequent plantings'.

3 PROS AND CONS

Terminator has some potential benefits. For one thing, it could allay one of the concerns of some opponents of genetically modified crops, which is the risk that genes from these plants may cross over to other species, a phenomenon called horizontal gene transfer that, ironically, many advocates of GM agriculture dismiss as being nothing to worry about anyway. In addition, secure protection might encourage further investment in agricultural biotechnology and plant breeding including in directions that benefit small farmers. This is a serious matter. To date, far too little welfare-enhancing scientific research is targeted at the poor, who tend to find that scientific revolutions tend to pass them by.

The former potential benefit is certainly plausible. However, the latter is, at this stage, completely speculative. There is some anecdotal evidence that weak plant variety protection in the US, which until quite recently allowed 'brown-bagging' (the sale of harvested seed by farmers), led to the closing down of at least one commercial seed production programme (Srinivasan and Thirtle 2002). But this was before the law was tightened up in the mid-1990s and when the patenting of plants was less common than it is today. Besides, if their public justifications are anything to go by, the USDA and Delta and Pine Land are more interested in applying terminator in foreign markets than the domestic one. Swanson and Goeschl support a more sceptical view when they note that 'the 40 year long experience with hybrid (use restriction) technologies is one of enhanced rent appropriation *but little change in investment patterns*' (emphasis added) (2004). The authors explain this phenomenon on the basis that so much of the public sector research investments during this time have been targeted at the improvement of non-hybrid crops.

Perhaps the main problem with terminator, if we suppose for a moment that the technology *will* encourage small farmer-oriented research, is its restriction on seed replanting, exchange, diffusion and on-farm breeding activities. To explain why, it is important to understand that developing country subsistence farmers generally acquire their seeds from their own farms or those of neighbours. Maintaining the freedom to do this is very important for two reasons. First, subsistence farmers often lack funds or credit to buy seed at the start of each planting season. For them, buying seed is a considerable investment. If it turns out that the benefits of terminator-protected seed are insufficient to compensate for its higher price, farming may become even more risky for the poor. Admittedly, farmers can presumably return to their traditional varieties, but one poor harvest accompanied by increased debt may be enough to cause destitution. Second, many developing country small-scale farmers do much more than simply grow seed produced elsewhere. Indeed, local varieties are themselves the result of generations of improvement through on-farm selection and experimentation, and nowadays such practices can involve modern varieties that may need to be adapted to suit local conditions. Turning such farmers into mere customers of companies selling terminator-protected seed will halt such practices. This may not only be detrimental to local food security but, if it became a global phenomenon, could weaken plant breeding efforts worldwide by reducing the variety of germplasm available.

For these reasons, subsistence farmers are unlikely to be sympathetic to, or gain much comfort from, Collins's assertion that 'the centuries-old practice of farmer-saved seed is really a gross disadvantage to Third World farmers who inadvertently become locked into obsolete varieties because of their taking the "easy road" and not planting newer, more productive varieties'.[6] If yields from GURT-protected seed prove to be disappointing, and if such seed is more dependent on inputs like agrochemicals than traditional varieties (which is often the case with modern varieties), then farming communities could suffer destitution. Defenders will no doubt argue that farmers still have a choice; that they can simply return to their traditional varieties if they prefer to avoid GURT-protected seed. But certain policies may erode their freedom to choose what they can plant in their fields. In many developing countries, government support for farmers including credit is sometimes made conditional on the planting of particular crops and types of seed, such as hybrids. Also, seed aid may be used by providers as a way to promote the use of particular crops and seeds. In addition, seed regulations in some countries require farmers to sow seed from an official list of approved varieties.

4 CONCLUSION: TERMINATOR AND THE PATENT SYSTEM

The widespread adoption of terminator technology, something that denial of a patent could ironically hasten, is unlikely to be the disaster that its strongest critics claim. Undeniably, it is a rational solution to the failure of existing IPRs and GURTs such as hybrids to ensure that breeders secure profitable returns on their investments in more than just a few crop species. Nonetheless, there are good reasons to be concerned that it may weaken plant breeding efforts worldwide by reducing the variety of germplasm available, undermine public sector research targeted at, and sometimes carried out *with*, poor farmers, and threaten the freedom that millions of farmers depend on to acquire their seed for free or at very low cost. Indeed, terminator technology has the genuine potential to seriously disrupt poor world agricultural systems that support the livelihoods of hundreds of millions of people.

For many critics, being able to patent such a technology is an indictment of the patent system. One may indeed reasonably question whether or not society should be encouraging such research through the promise of a patent monopoly. Moreover, it is legitimate to be concerned that protecting seeds through both patents and GURTs is overprotective in a similar way that support for encryption through copyright law in the form of banning circumvention devices is overly generous to owners.[7] In addition, countries might well consider such inventions to be immoral or contrary to *ordre public*, and this is a decision that deserves to be respected. Legally, they are on secure ground since they have the right to determine their own criteria for what is immoral or contrary to *ordre public*. India has already banned terminator and one can easily envisage other developing countries following suit. For such countries, terminator is clearly not the sort of technology they want their patent systems to encourage.

However, it seems to the present author that there is no particular need to respond to terminator-type patents by broadening the application of these exceptions.[8] In fact, if terminator could not be patented on these or other grounds, this might encourage research in this area even more. After all, GURTs would appear to be especially useful in jurisdictions where IPR protection is weak. Besides, the patent itself is not a right to commercially exploit. The freedom to use such technologies should not be automatic but subjected to an approval process founded on sound science, socio-economic and environmental assessments of its impacts and, arguably, the precautionary principle. Patent-granting authorities are not qualified to do any of this.

There are two main issues highlighted by the terminator patent, even though these have often been ignored in the sound and fury. First, terminator technology exemplifies the way agricultural research is more and more expensive, commercially oriented and technologically advanced. The consequence of this is that the sector is becoming one in which an ever smaller number of companies is able to enter it, while those that are already in it and can compete come to dominate it. In fact, terminator may accelerate this process of corporate concentration, which is already quite noticeable,[9] while further undermining public sector research. It may do this by tightening the locks on plant genetic resources so others must either do without them or pay licence fees that might prove too financially burdensome for competitors or potential competitors and public sector institutions (Swanson and Goeschl 2002). If so, countries need to adopt competition and seed regulations that ensure that farmers and consumers continue to have a choice and that maintain the public sector's freedom to operate in agricultural research. One possible measure they might consider is compulsory licensing. However, one should bear in mind that if companies cannot easily capture the benefits from technological innovation through the patent system, this may make them even more determined to control markets through other means, such as by taking over supply networks and by integrating horizontally so they can market 'packages' of products that need to be used together (Rangnekar 2002).[10] This issue is difficult to resolve but at least forces us to reflect upon what should be considered an appropriate rate of return on private investments.

Second, we allow the private sector to monopolize agricultural research at our peril, and the peril especially of developing country farmers, who are bound to be ignored in the same way that drug companies ignore the diseases of the poor for sound economic reasons. Terminator could even make the situation worse for the poor because even if GURT-protected seeds are developed for the use of poor farmers, this may backfire on them because the scope for on-farm breeder experimentation, which is often necessary to adapt varieties so they better meet the specific needs of farmers, will be reduced. It is also a very bad thing if public sector researchers in other countries have the same motivations as the USDA in supporting terminator. Public sector agricultural research is declining worldwide (Knight 2003). Yet research targeted at poor farmers is as necessary as it has ever been. Not all public sector research does this, as the terminator research amply demonstrates, but we can be sure that even less will be done if business is left to conduct all the research. Indeed, the termination of public sector research may be a bigger problem for poor farmers than terminator technology. If we allow this to happen, we may well prove the Devil was right about us after all.

NOTES

1. Since renamed Action Group on Erosion, Technology and Concentration (ETC Group).
2. US Patent No. 5 723 765 (issued 3 March 1998) (Control of Plant Gene Expression).

3. In a discussion on private appropriation in the business of plant breeding, Rangnekar (2002) identifies four methods employed. These are (1) IPRs and seed market regulation; (2) organizational solutions; (3) discontinuous heritability; and (4) planned obsolescence. He classes hybrid and terminator technologies as methods of discontinuous heritability.
4. Growing for seed is more costly but, on the other hand, seed prices are significantly higher than grain prices (Rangnekar, personal comment 2003).
5. The correspondence between the United States Patent and Trademark Office and the Office of the Union and the US government's response to the memorandum is incorporated in UPOV document CAJ/7/7 as Annex II.
6. Transcript of presentation made at side event of the fourth meeting of the Conference of the Parties to the Convention on Biological Diversity at Bratislava, Slovakia in June 1998 (on file with author).
7. Art. 8 of the 1996 WIPO Copyright Treaty and Arts. 10 and 14 of the 1996 WIPO Performances and Phonograms Treaty.
8. This is not to deny the possible existence of other grounds for doing this.
9. According to the activist group, Action Group on Erosion, Technology and Concentration, 'the top 10 seed companies control approximately 30% of the $24.4 million commercial seed markets worldwide' (ETC Group 2001, p. 9). Tansey notes that 'the US seed industry, once the preserve of many small firms, has become dominated by five major firms – in part as a response to litigation over broad patents awarded in the early days of GM in the USA' (Tansey 2002).
10. As early as 1987, it was noted that 'a substantial amount of plant research in private firms has been aimed at developing various types of seed-chemical packages that reinforce rather than threaten sales of agricultural chemicals' (Buttel and Belsky 1987). For example, Monsanto developed and patented transgenic soybeans, canola, cotton and corn containing a gene providing resistance to its Roundup (glyphosate) herbicides. Monsanto's patents protect the gene for Roundup resistance and all plants containing it, and these have several more years to run. As farmers who buy these 'Roundup Ready' seeds are contractually obliged to purchase Monsanto's patented herbicides, sales of the seeds are good for sales of the herbicides and vice versa. It is unclear, however, that this strategy will work in the long term. Roundup Ultra went off patent in 2000 and farmers may well turn to cheaper versions sold by competitors (Dutfield 2003, p. 149).

REFERENCES

Berlan, J.P. and R.C. Lewontin (1998), 'Cashing in on Life – Operation Terminator', *Le Monde Diplomatique*, December, http://www.mondediplo.com/1998/12/02gen

Bugos, G.E. and D.J. Kevles (1992), 'Plants as Intellectual Property: American Practice, Law, and Policy in a World Context', *Osiris*, **7**, 75–104.

Buttel, F.H. and J. Belsky (1987), 'Biotechnology, Plant Breeding, and Intellectual Property: Social and Ethical Dimensions', *Science, Technology, and Human Values*, **12**(1), 31–49.

Collins, H. (1998), Transcript of presentation made at side event of the fourth meeting of the 'Conference of the Parties to the Convention on Biological Diversity at Bratislava', Slovakia, June (on file with author).

Convention on Biological Diversity (CBD), Rio de Janeiro (Brazil), 5 June 1992 (entered into force 29 December) 1993 (1992) *International Legal Materials*, **31**, 818.

Dutfield, Graham (2003), *Intellectual Property Rights and the Life Science Industries: A Twentieth Century History*, Aldershot: Ashgate.

ETC Group (2001), 'Globalization, Inc. Concentration in Corporate Power: The Unmentioned Agenda', *ETC Group Communiqué* No. 71.

Knight, J. (2003), 'Crop Improvement: A Dying Breed', *Nature*, **421**, 568–70.

Rangnekar, D. (2002), 'R&D Appropriability and Planned Obsolescence: Empirical Evidence from Wheat Breeding in the UK (1960–1995)', *Industrial and Corporate Change*, **11**(5), 1011–29.

Rural Advancement Foundation International (RAFI) (1998), 'US Patent on New Genetic Technology Will Prevent Farmers from Saving Seed', RAFI press release dated 11 March.

Sehgal, S. (1996), 'IPR Driven Restructuring of the Seed Industry', *Biotechnology and Development Monitor*, **29**, 18–21.

Shaw, G.B. (1903) [2000], *Man and Superman*, London: Penguin Books, p. 142.

Srinivasan, C.S. and C. Thirtle (2002), 'Impact of Terminator Technologies in Developing Countries: A Framework for Economic Analysis', in T.M. Swanson (ed.) *Biotechnology, Agriculture and the Developing World: The Distributional Implications of Technological Change*, Cheltenham, UK and Northampton, MA, USA: Edward Elgar, pp. 150–76.

Swanson, T.M. and T. Goeschl (2002), 'The Impact of GURTs: Agricultural R&D and Appropriation Mechanisms', in T.M. Swanson (ed.) *Biotechnology, Agriculture and the Developing World: The Distributional Implications of Technological Change*, Cheltenham, UK and Northampton, MA, USA: Edward Elgar, pp. 44–66.

Swanson, T. and T. Goeschl (2004), 'The Impacts on Poor Countries of Technological Enforcement Within the Biotechnology Sector'. Unpublished manuscript on file with author.

Tansey, G. (2002), 'TRIPS with Everything? Intellectual Property in the Farming World', Southwell: Food Ethics Council.

UPOV (2003a), Position of the International Union for the Protection of New Varieties of Plants (UPOV) concerning Decision VI/5 of the Conference of the Parties to the Convention on Biological Diversity. Communicated to the Secretariat of the CBD. Published on the internet at: http://www.upov.int/en/about/pdf/gurts_11april2003.pdf.

UPOV (2003b), Memorandum prepared by the Office of the UPOV on genetic use restriction technologies. Document prepared by the Office of the Union [CAJ/47/7]. Published on the internet at: http://www.eldis.org/static/DOC11879.htm.

11. Patents, patients and consent: exploring the interface between regulation and innovation regimes

Graeme Laurie

1 INTRODUCTION

The involvement of patients in medical research is a practice as old as the discipline of medicine itself, and the role of consent in this practice has come under particularly close scrutiny in the last half-century as cultural and societal attitudes towards science, medicine, health professionals and the rights of individuals have undergone radical re-appraisal. Policy-makers have come to rely heavily on the concept of 'informed consent' as a legal and ethical panacea when dealing with new dilemmas in modern medicine. This has extended even to the realm of intellectual property law, as demonstrated by the attempt – albeit failed – to introduce an informed consent requirement into the provisions of Directive 98/44/EC on the Legal Protection of Biotechnological Inventions (EC Directive 98/44/EC 1998).

Notwithstanding such developments, the legal requirements proscribing (1) what research subjects should be told in order for their consent to be valid, and, perhaps more important still (2) which legal remedies are available if valid consent is not obtained, remain contested. Ethical debate has far outstripped legal developments in addressing these questions. This chapter examines the legal basis of informed consent doctrine as a means to respect patients and persons who act as research subjects and it considers the extent to which consent to patenting does, or should, play a role in contemporary European patent law. The conclusion is that while a limited role for consent may be identified within the patent regime – which in turn requires us to revisit the delicate relationship between patent and regulatory regimes – this is not a complete policy, nor is it necessarily the best policy to protect or advance the interests of patients as research subjects; far less does it advance their interests in the commercialization of the products of the research in which they have participated. While the chapter focusses on developments in Europe, it reveals common values of universal application about the importance and significance of consent; it also raises broader questions about the interaction of patent systems with other social systems.

2 THE ROLE OF CONSENT IN TREATMENT AND RESEARCH

There is a variety of different ways that individuals can be involved in research. For example, research can be done (1) on persons; (2) with material taken from persons, such as tissue or blood samples, and (3) with information about persons (whether or not derived from material from those persons). In the vast majority of jurisdictions and circumstances, however, the consent of these persons is required, both legally and ethically, to legitimate the research process. Indeed, ethical and legal sensitivities towards the informational needs of research subjects have developed considerably in recent years. The legal trend over the last two decades has been towards more specific, detailed disclosures relating both to treatment and research decisions. The standard to be met is that of informed consent. And, while some have pointed to the inherent limits of informed consent – for example, highlighting that specific, comprehensive informedness is simply an impossible standard to achieve (for example O'Neill 2002) – and others have argued that a criterion of informed consent is not always appropriate for many contemporary (longitudinal) research projects,[1] the metaphorical power of consent is now overwhelming, and its reach into regulatory regimes is unprecedented. An example is the Directive 2001/20/EC on the Conduct of Clinical Trials on Medicinal Products for Human Use (EC Directive 2001/20/EC 2001), which was due to be implemented by all Member States by May 2004. This seeks to impose a uniform consent requirement across the Union. It defines informed consent as '[a] decision, which must be written, dated and signed, to take part in a clinical trial, freely after being duly informed of its nature, significance, implications and risks' (EC Directive 2001/20/EC 2001, Article 2 (j)).

This embodies the essence of *informed* consent: the participant should receive enough information in order to take a meaningful decision about the proposed research (or treatment as may be). This, however, then begs the obvious question: what is meant by the 'nature, significance, implications and risks'? It also raises the issue of the scope of the obligation to disclose: what, if anything, should research participants be told *beyond* the immediate risks or consequences for their person? It is this question that is the focus of this chapter. More especially, we consider whether a principled and ethically sound commitment to informed consent includes a requirement to disclose future commercialization plans, including the prospect of patenting inventions developed as a result of individuals' participation in research. But in order to determine what should be disclosed we must begin with the logically prior question: what function does consent serve?

3 WHAT IS CONSENT FOR?

Consent is a means of showing respect for persons and for their fundamental rights, such as the right to self-determination, the right to bodily integrity, or the right to respect for privacy. Consent is a means to legitimate an otherwise unacceptable act, such as an unauthorized touching of the body or an unwarranted use of personal information. Consent, therefore, is not an end in itself, but rather a means to an end – it is a means to respect people. A number of consequences flow from this.

First, it indicates that the fundamental value at stake is not the obtaining of consent, but rather the furnishing of respect. This, in turn, implies that consent may not be the only means of showing respect, and by extension that consent may not always be required. Numerous applications of this can be found in the research setting. For example, we often forego consent in studies on incapacitated persons precisely because these persons cannot give meaningful consent.[2] To adhere too stringently to the consent model would mean that research into the conditions from which these persons suffer would not otherwise be done. In the context of personal information, we often forego the need to obtain consent because of the impracticality of seeking it or because the fundamental interests at stake, namely privacy, can be sufficiently protected by other means, such as anonymization of the personal data. The justification often given here is that a significant public good is served by allowing research to proceed.[3] Similar reasoning has been applied to justify research without consent on previously donated samples (Medical Research Council 2001, Paras. 3.5 and 10.2). In each of these cases we find other means of protecting and respecting the persons in question, such as requiring rigorous ethical analysis of the research proposal, restricting the scope of the research to precisely defined areas (for example, only permitting research into conditions from which the research participants are suffering), guaranteeing anonymity and/or confidentiality, and ensuring periodic review of the entire research programme.

In other circumstances it is self-evident that consent alone cannot absolve researchers and others from broader obligations to respect and protect research subjects. An apposite and disturbing example of this is the recent failed drug trial which took place at London's Northwick Park

Hospital in March 2006 when six healthy volunteers experienced extreme auto-immune reactions after the administration of an experimental monoclonal antibody, TGN1412. The product had previously been tested on primates with no evidence to suggest heightened risks in humans, but questions have since been raised as to the robustness and appropriateness of such pre-trial procedures, as well as concerns about the conduct of the trial itself. A final report from the Medicines and Healthcare Products Regulatory Agency (MHRA) in May 2006 stated that: 'This investigation indicates that the adverse incidents did not involve errors in the manufacture of TGN1412 or in its formulation, dilution or administration to trial participants'.[4] Notwithstanding, the report acknowledges that this is a relatively new form of biological drug and the incident raises significant scientific and medical questions about the risks involved and what research subjects should be allowed to consent to. The matters were referred to an expert scientific group established by the Secretary of State for Health for further consideration.

None the less, the spectre of the individual's right to self-determination as exercised through the right to consent or to refuse has become a powerful regulating force in the relationship between health care professionals and patients, and, in turn, between researchers and patients as research subjects. Numerous international and national legal instruments give preeminence to the need for consent as the ultimate mark of respect for the individual.[5]

The paradigmatic legal remedy for failure to obtain informed consent, domestically at least, is first and foremost the action of assault, for which damages are payable for any unauthorized touching of the body. This, of course, is of utility only when the treatment or research in question involves physical contact. An action in negligence might also lie if it can be shown that a duty of care was incumbent on another party to obtain valid informed consent, that this was not obtained, and that a recognized form of harm arose as a result.[6] Once again, however, the harm here is usually of a physical kind, such as the manifestation of a risk in care or research that – had the individual been told – would have led her to refuse the procedure.[7] Far less clear is what should happen when individuals have been disrespected in other ways, such as through the misuse of their personal data (which may find a remedy in privacy laws, where these exist), or through unauthorized use of tissue samples (which rarely finds a specific remedy for the individual in question).[8] In such circumstances, regulatory regimes may intervene to withdraw regulatory approval for research, for example, or to discipline aberrant health care professionals for professional misconduct. Consent per se offers little in the way of direct legal remedies for the individuals who feel aggrieved.

Much of the power of consent lies in the perception that consent is empowering, that is, that it makes the patient or research participant a more equal partner in his or her relationship with the health care professional or researcher. For many, consent renders the adage 'Information is Power' a truism. This belief, however, is subject to two fundamentally important qualifications. First, it must be recognized that the only power that consent furnishes is the power to refuse. While this is self-evidently helpful in ensuring that no one is enrolled in research or subjected to treatment against their will, it provides no possibility of influence for those who are willing to participate but have reservations about the treatment or the conduct of the research, or, indeed, about other aspects of the process such as commercialization. Medical treatment models largely perpetuate the power of the health care professional, while regulatory models for research are not currently set up to allow for negotiation of terms: participation is often an all or nothing affair from the perspective of the prospective participant.[9]

Consent is usually a one-off event that does not provide for any continuing control or possibility of influence on the part of individuals beyond complete withdrawal from the treatment or project at some future time.[10] This brings us to the second qualification: a consent model is invariably blind to the power imbalance that exists in all relationships between doctors and patients and between researchers and research participants. Furthermore, to the extent that individuals have any power at all to make informed decisions, this is entirely contingent on what they are told, and this in turn places all of the power – at least in the first instance – in the hands of the professionals. We will return to these limitations of consent towards the end of this chapter. For present purposes, they do not detract from the current popularity enjoyed by the concept among policy- and law-makers, nor have they restricted the reach of consent into many new areas of life. The crucial issue in legal terms is what are people *entitled* to be told in any given context? That is, what is the nature and extent of the obligation to disclose?

4 WHAT SHOULD PEOPLE BE TOLD?

Legal standards of disclosure vary across jurisdictions and depend on whether the relationship is one of doctor/patient or researcher/subject. In the United Kingdom, for example, an assault action will not lie, in the absence of fraud, if a patient is informed in 'broad terms' about the 'nature and purpose of a medical treatment procedure' (*Chatterton v Gerson* 1981). In general, this receives a restrictive interpretation, leaving little scope for a successful action. Two particular hurdles are faced in the context of a negligence action: first, it must be shown that the harm complained of was caused by the negligent failure to disclose, and when that harm is not physical harm the categories of recovery are extremely limited. If, for example, the claim is mental anguish, then the British courts have tended to require actual psychiatric disturbance, as opposed to mere affront to dignity or integrity (although see most recently Chester v Afshar 2004). Second, until very recently, and perhaps even still in some cases, the medical profession has been the judge of what should be disclosed to patients.[11] This is slowly changing from a professional-oriented test to a more (objective) patient-oriented test, but still it begs the question of what the objective patient would wish to know (*Pearce v United Bristol Healthcare NHS Trust* 1998). Ultimately, this can only be a matter for the courts to decide, influenced, as always, by public policy considerations.

These rules have been developed, however, in the context of medical care and treatment. In the research context the standard of disclosure should be different because different considerations apply: the kind of relationship that exists between the parties is not the same as between doctor and patient, the primary aim of the relationship is not the patient's health but rather the generation of new generalizable knowledge, and the possibility of paternalistic behaviour towards the patient in his or her own best interests is entirely excluded. In most jurisdictions this translates to a standard of disclosure that suggests that the subjective needs of the subject should dictate the level of information to be communicated.[12] This, in turn, focusses on the materiality of the information to the decision of the individual: how central or influential would a piece of information be to the decision-maker such that its disclosure is required to allow them to make an informed decision to their own satisfaction? Put more simply, would the person consider it important to weigh up a particular piece of information in coming to a decision? Might this information make a difference to their decision one way or another? The information that concerns us here is the prospect of commercialization of the results of research, or more particularly, the prospect of patenting inventions that emerge from that research.

It is undeniable that the commercialization of research is perceived with concern in many quarters. Numerous public engagement studies indicate that public confidence is shaken by the prospect of private profit from public involvement in biotechnological research, and most especially by the prospect of patents over the products of that research (see, for example, Cragg Ross Dawson 2000, and Hapgood et al. 2004). There is, moreover, a definite trend towards requiring fuller disclosure of commercial interests on the part of researchers. For example, the UK Medical Research Council requires that this forms part of consent procedures for research sponsored by the Council and British Local Research Ethics Committees increasingly require that such disclosures be made (Central Office for Research Ethics Committees).

In the infamous (and overly debated) case of *Moore v Regents of the University of California* (1990), the Supreme Court of California held in favour of the plaintiff for the failure of his physician to obtain informed consent and for the breach of fiduciary trust occasioned when the clinician omitted to disclose his commercial interests in Mr Moore's cells – excised during a spleenectomy – which were used to develop an immortal cell-line that was ultimately the subject of a successful patent application. Of most significance for present purposes, however, was the fact that while Moore won his case and received damages for lack of consent and breach of trust, there was no suggestion that the affront done to him could, or would, in any way affect the validity of the patent. In truth, there is no mechanism for this to occur in the United States. But the case reveals a deeper malaise for all patent systems and research regulatory systems about the extent of our commitment to respecting individuals; it speaks of the consequences, in particular, of using consent as a mechanism to furnish that respect; and it leaves us to question the extent to which consent – or the lack of it – should impact on the operation of the patent system. These are the questions that concern the remainder of this chapter.

5 CONSENT IN PATENT LAW: FROM ETHICAL PRINCIPLE TO LEGAL UNCERTAINTY

What can be said of the ethical and legal position in respect of consent to patenting in the research context? Is this part of the obligation of disclosure of researchers to research subjects? The European Commission's own Group of Advisers on the Ethical Implications of Biotechnology (now re-named the European Group on Ethics in Science and New Technologies) certainly thought so in 1996 when it opined that:

> The ethical principle of informed and free consent of the person from whom retrievals are performed must be respected. This principle includes that *the information of this person is complete and specific, in particular on the potential patent application on the invention which could be made from the use of this element*. An invention based on the use of elements of human origin, having been retrieved without respecting the principle of consent, will not fulfil the ethical requirements.[13] (emphasis added)

This approach was picked up by the European Parliament in its numerous debates on the passage of the EC Directive on the Legal Protection of Biotechnological Inventions. Indeed, such a requirement formed one of the 66 amendments to the Directive that were proposed in 1997. It was also the only amendment not to survive the process. Accordingly, no mention of consent in these terms is to be found in the articles of the 1998 Directive. Recital 26, however, states: 'Whereas if an invention is based on biological material of human origin or if it uses such material, where a patent application is filed, the person from whose body the material is taken must have had an opportunity of expressing free and informed consent thereto, in accordance with national law'.

Two elements relating to this Recital have been disputed, namely, its legal status and its meaning. On the first point, it is strongly arguable that the location of such a provision in the recitals makes it, at best, an aid to the interpretation of the articles themselves, and most notably, the morality provisions in Article 6, which we will consider presently. It would follow, then, that there is no obligation as such to implement these consent provisions expressly in domestic law.[14] Second, dubiety remains about the true meaning of the Recital: must the consent be to the *filing* of the patent application or to the *taking* of the material from the participant's body? The French translation is equally ambiguous; the Spanish less so, speaking of 'the person from whom the samples were taken should have the opportunity to give his free consent with the correct information about the said samples'.[15] Thus, this may be no more than a simple restatement of the requirement to ensure that individuals consent to taking part in research, which is a far cry from a requirement of consent to the filing of a patent application. Some have argued, however, that given the general acceptance of the rule that non-consensual taking of samples is unlawful, such a measure is redundant on this reading, and 'the more robust, literal and . . . unambiguous interpretation . . . is that free and informed consent must be given not only to the taking of the tissue but also to the use of the tissue in research work that might lead to an application for a patent' (Beyleveld and Brownsword 2001, p. 203).

Such an approach to the interpretation of European patent law is, in fact, reflected in the tenor of recent discussions in the European Patent Office (EPO) about the meanings of Article 6 and associated recitals. The Opposition Division, for example, has advocated a claim of redundancy to support a broad interpretation of European patent law as this relates to patents involving uses of human embryos. In the *Edinburgh Patent* case,[16] the Division was called upon to interpret Article 53(a) of the European Patent Convention (EPC) and Rule 23(d)(c) EPC, following which European patents shall not be granted for 'uses of human embryos for industrial or commercial purposes'.[17] The Division noted that the provision could be interpreted in two ways: narrowly, to mean that only commercial uses of human embryos *as such* are excluded from patentability, or broadly, to mean that human embryonic stem cells – which can only be obtained by destroying an embryo – are also unpatentable. The Division preferred the latter approach, arguing that since embryos *as such* are already protected by Rule 23(e),[18] a similar interpretation of Rule 23(d)(c) would be redundancy and this could not have been the intention of the legislator. Such a broad approach has most recently been followed by an Examining Division of the EPO in another embryonic stem cell application,[19] and to the extent that these rulings represent any form of precedent, it would suggest that the trend should be towards a similar interpretation of other patent law provisions to obviate redundant outcomes. If so, it would follow that the consent provisions should similarly be read to mean more than merely consent to participation in research (which is already well provided for through regulation law), and that they might indeed extend to the need to obtain consent to the pursuit of patent applications (being the most materially relevant factor for research participants vis-à-vis the application of patent law to their circumstances).

6 THE DUTCH CHALLENGE TO THE BIOTECHNOLOGY DIRECTIVE

Concern about the consent provisions in the EC Directive was a central feature of the Dutch challenge to the instrument in 2001 (*Kingdom of the Netherlands v Council of the European Union and the European Parliament*). The Netherlands, joined by Italy and Norway, argued that the fundamental right to self-determination was breached by the 'absence of a provision requiring verification of the consent of the donor or recipient'. This, however, addressed only the question of the legal status of consent provisions in the Directive and not their meaning. None the less, Advocate General Jacobs endorsed the Dutch view in his Opinion when he stated: 'The fact that there is nothing in the body of the Directive ensuring that human material is managed carefully . . . [through consent] . . . must be considered contrary to fundamental right . . . [of human integrity and self-determination]'.[20]

The European Court of Justice (ECJ) disagreed. It chose to see the patent regime as a system apart – concerned only with the grant of a patent. The reach of this system did 'not . . . extend to activities before and after that grant, whether they involve research or the use of the patented products'. As support, the ECJ claimed that:

> [t]he grant of a patent does not preclude legal limitations or prohibitions applying to research . . . [t]he purpose of the Directive is not to replace the restrictive provisions which guarantee, outside the scope of the Directive, compliance with certain ethical rules which include the right to self-determination by informed consent. (*Kingdom of the Netherlands v Council of the European Union and the European Parliament* (2002), Para. 80)

Accordingly, the Directive was not deemed insufficient for lack of direct commitment to the principle of requiring informed consent (broadly defined).

7 WHITHER CONSENT IN PATENT LAW?

This ruling confirms, then, that informed consent is not a direct part of European patent law, and the kernel of the argument to support this relates to the appropriateness of relying on the patent system to discharge functions that are otherwise regulatory. Expressed thus, the ratio has much merit, for the need to maintain a clear division of labour between the two systems is axiomatic. But it does not follow that there is no relationship between these regimes, nor, indeed, that they do not influence each other. None the less, the decision does reflect the historical tendency to view the patent system as an entity in isolation, driven only by its own internal dynamics and unencumbered by broader social agenda.[21] This attitude has been changing in recent years,[22] but we are only now beginning to explore the nature and extent of the relationships that the patent system has with other social systems. Moreover, even if we accept the ECJ's starting premise that the patent system should not usurp the regulatory system, this does not require that we also accept that there is no role for consent within the former system. Indeed, why should the patent system not reflect the regulatory system in its commitment to respect individuals who participate in (medical) research?

In order that the discussion proceeds meaningfully from here, we must distinguish between three different concepts. First, we have *usurpation* of the function of one system by another, that is, the inappropriate supplanting by one system of functions better discharged by another. Second, we can consider *redundancy* of function, that is, the unnecessary repetition of provisions in one system that already appear, and function adequately within, another. We can ask, for example, whether consent provisions in patent law would add anything that is not already encompassed by the regulatory regime? Finally, we can contemplate the *congruence* of systems such that they reflect common ends or shared values of a society or social group in a manner that is consistent across the systems in question. An example would be the fundamental value of the principle of respect for persons, which is common both to regulation law and patent law.[23]

The ECJ engaged only with the first of these concepts – usurpation – and its conclusion that the patent system should not encroach unduly on the regulatory system is undoubtedly correct. But the ECJ ruling does not say that the articles of the Biotechnology Directive cannot be interpreted to include a requirement of informed consent to the filing of the patent. In fact, this might indeed be desirable from a congruence perspective depending on what it would mean for research participants and the patent system, and subject, of course, to a challenge of redundancy. If we suppose for the sake of argument that this is a reasonable interpretation of the Directive – and one that it is open to Member States to adopt[24] – we can ask how the position might be defended and how such a requirement might work in practice. In other words, we can enquire:

- Is there, in principle, a distinct and defensible role for consent in patent law?
- If so, how might it work in practice?

8 PRINCIPLES

The patent and regulatory systems cannot be treated as mutually exclusive because a relationship necessarily exists between them by virtue of their nature, content and the consequences of their operation. The potential exists for one to influence the other and the challenge is to understand these dynamics and to determine how and whether changes can, or should, be effected. But these are complex systems and this is not an appropriate place to attempt a detailed analysis.[25] None the less, the example of consent is apposite to explore the interface in both principle and practice. From a principled point of view we can ask, then, what might informed consent to patent filing bring to either system, and does such a proposal stand up to analyse from each of the perspectives of (1) usurpation, (2) redundancy, and (3) congruence?

8.1 Would Consent Provisions in Patent Law Usurp the Regulatory Regime?

The answer to this question depends on the content of the consent provisions under consideration. The Dutch challenge to the Biotechnology Directive took a twofold approach to the question of consent.[26] First, it advocated a control-based argument implying that a donor or recipient of material has a right to influence the fate of his or her body or parts thereof. Second, it advanced a paternalistic argument that the Directive failed to contain sufficient protection for patients who might otherwise receive biotechnologically manipulated material without their knowledge or consent. Neither of these, then, had anything directly to do with patenting, yet both were premised on the view – for which there is much existing legal support – that consent is an aspect of the fundamental right to integrity.[27] But consent to what? Advocate General Jacobs stated in his Opinion that: 'although the requirement of consent to *all* potential uses of human material may be regarded as fundamental, patent law is not the appropriate framework for the imposition and monitoring of such a requirement'[28] (emphasis added).

This must be true if the question is whether patent law should (or can) regulate *all* such uses. For one, it simply cannot do so given that the remit of any patent regime is solely the grant or revocation of a market monopoly. It has no reach into the fields of ethical review, scientific validity, public safety or a multiplicity of other areas better handled by regulation. Any such attempt would indeed be a usurpation of the latter by the former. But the extent to which overlap between the systems should be tolerated is another matter. The mistake of the Dutch challenge was to ask too much of the patent system and to claim too much for the role of consent provisions in patent law. In contrast, there is no reason to suppose that a suitably drafted consent requirement in patent law – defined as a specific requirement to demonstrate prior consent to patent filing – could in any way supplant the myriad other consent requirements under a regulatory regime. Indeed, this is a logical and temporal impossibility given the nature of the innovation trajectory taken by most inventions. Regulatory ethical requirements concerning participants must be met before research begins, while questions of compliance with consent provisions relating to patent applications would only arise when patent protection was sought, normally well after direct research subject participation has ended. So, a properly and narrowly defined consent provision could survive a charge of usurpation. This brings us to challenges of far more import, namely, those based on a claim of redundancy.

8.2 Avoiding Redundancy (or Achieving Additionality)

Issues of redundancy were raised earlier to support the argument that Recital 26 of the Biotechnology Directive should be interpreted to mean consent to the filing of a patent application and not merely consent to involvement in research. None the less, even on this analysis it might be argued that there remains no reason to give effect to such a provision in patent law because this would still represent redundancy: the regulatory regime already requires informed consent to participate in research and the standard to be reached is that of (subjective) materiality. This disclosure standard could easily incorporate a requirement to disclose financial interests on the part of researchers or future commercialization intentions, including the intention to seek patent approval. If this standard is not met, ethical approval will not be granted, or if it has been granted, ongoing research should be stopped when the anomaly comes to light. Moreover, civil remedies would be available for aggrieved research subjects as discussed above, and indeed this was precisely the position that applied in *Moore*. But, as that case also demonstrates, if there is no connection between the regulation and patent systems such remedies leave a patent on related research products unaffected – a patent that facilitates commercial exploitation of an invention that has been developed through unacceptable practices and unauthorized instrumentalization of research subjects. This suggests that the claim of redundancy may not be well founded in these circumstances. To link regulatory requirements of informed consent to the validity of a patent could, in fact, provide an additional level of protection, achieved through the threat of patent denial or revocation lest the requisite consent has not been obtained. The patent system, we are so often told, is driven by incentive. If this is true, cannot the threat of patent denial or revocation act as a disincentive to unethical conduct? This would be a forceful message about the unacceptability of certain practices and it might have a positive influence on any such practices for the future. Even if the patentee is not the researcher or person who obtains consent, it is in each party's interests to obtain that consent, assuming a professional, economic relationship between them. And although the mere denial of a patent does not represent any direct remedy for those research participants who have been disrespected, it serves both as a sanction against those who have disrespected them and ensures that the key object of concern – the economic advantage represented by a patent – is thwarted.

A more worrying aspect, however, is that patent denial or revocation cannot prevent use of the material obtained from research participants. Does this, then, fundamentally undermine the approach? Arguably not. Rather, it highlights the limitations of the patent system in *regulating* science and scientific conduct. It should, instead, re-enforce the message that while overlap exists between the systems, the patent system, for its part, has a restricted role to play. This does not mean, however, that it cannot bring something valuable to the overall approbation of the research trajectory, so long as usurpation is avoided and additionality is achieved. The contribution is not in the realm of direct regulation, but in *incentivization*, or perhaps more accurately, *disincentivization*.

8.3 Achieving Congruence

The patent system does not exist in a moral vacuum. This is reflected in the wording of Article 6 of the Biotechnology Directive, which states:

'Inventions shall be considered unpatentable when their commercial exploitation would be contrary to ordre public or morality; however, exploitation shall not be deemed to be so contrary merely because it is prohibited by law or regulation' (EC Directive 98/44 1998). The morality provisions exist as a mechanism to register opprobrium at the prospect of commercial exploitation of particular inventions. Exploitation is precisely the concern at the heart of consent-based objections to commercialization practices in research – 'had we been told we would (or might) have refused' – and the commercial exploitation of inventions developed through practices that disrespect research participants is therefore also highly questionable.[29] According protection to these inventions compounds the affront. To the extent that we accept there is an overarching obligation to uphold the fundamental principles of respect for human dignity and the right to integrity,[30] the grant of a patent in the face of unethical research practices might be seen as a form of state-sanctioned unjust enrichment. This is an issue of seeking congruence between regulatory and patent regimes. Where the regulatory system has already mandated consent, and where a consent provision also appears in the patent law, congruence would suggest that each system should reflect and re-enforce this value preference in a manner that is true to the essentials of that preference. Thus, if consent is about ensuring that individuals can exercise their self-determination within a social framework that respects them and values their choices, one would expect disrespectful conduct to be sanctioned, or at least not supported by the state in contradictory fashion as between different segments of that framework. To do otherwise exposes these state-controlled systems to charges of inconsistency and, ultimately, challenges of legitimacy. While this is not to say that different factors and considerations do not apply in different contexts and circumstances,[31] for the patent system is not the regulation system nor vice versa, the clear and strong commitment in Europe to the well-accepted fundamental of human dignity and integrity across both of these systems points to a need for the congruence spoken of here. At the very least, it places the onus on Member States to justify examples of incongruence; something that has not happened to date to any acceptable degree.[32]

Finally, it is also noteworthy that the second part of Article 6(1) returns us to considerations of additionality in that it reveals a further reason to seek sanction from within the patent system itself. For, while it could be argued that the regulatory system is perfectly well equipped to impose commercial sanction on unethical researchers in the guise of licence refusal (effectively ruling out a domestic market), the ongoing availability of protection would remain an incentive to produce domestically under patent for exportation and exploitation abroad, as the EPO itself has pointed out.[33] This, once again, would merely perpetuate the initial affront.

9 PRACTICALITIES

Subject to the above qualifications, it is submitted that there is no compelling reason – in principle – why patent grant should not be considered as an integral part of the entire research approval process. In this sense, patent law can play its part in delivering judgement on the appropriateness of the innovation trajectory followed by any given invention, an essential part of which is the need to obtain informed consent from research participants. But how might this work in practice?

It has already been suggested that the morality provisions of European patent law are the obvious place to consider a role for consent. Until very recently,[34] however, these have been read very narrowly. Indeed, some have even spoken of a presumption of patentability (Llewelyn et al. 2000). Thus, in *HOWARD FLOREY/Relaxin* (1995) the Opposition Division (OD) of the European Patent Office (EPO) held that the prospect of patenting would have to 'be abhorrent to the overwhelming majority of the public' before the EPO would reject an application on morality grounds. This sets a high standard, but not impossibly so. Indeed, the Division ruled in the case that 'the patenting of DNA would indeed be abhorrent to the overwhelming majority of the public if it were true that the invention involved . . . an abuse of pregnant women' (*HOWARD FLOREY/Relaxin* 1995). The facts related to the patenting of H2 Relaxin, a naturally-occurring protein that is produced during childbirth to soften the pelvis and ease the passage of the child. Moral objections were raised to the invention but the patent was saved by the fact that the pregnant research subjects had consented. While it is unclear to what, exactly, that consent had been given – that is, whether it was to the *taking* of samples or to the subsequent *filing* of the patent application – the clear suggestion is that a lack of consent is a form of abuse, which in turn may be sufficient grounds to refuse or revoke a patent.

The rulings of the EPO are not, of course, binding on the individual Member States that apply the provisions of the Biotech Directive, nor, indeed, for that matter on the individual signatory states to the European Patent Convention (EPC). But the approximation of provisions in the Biotechnology Directive and the EPC is now so close – as a direct and deliberate policy[35] – that any Member State predisposed to challenge the presumption in favour of patentability could, it is submitted, rely on these judgements and the provisions of Article 6 to deny patent protection for lack of informed consent.

10 WHAT COUNTS AS IMMORALITY?

The *HOWARD FLOREY/Relaxin* ruling none the less perpetuates the confusion about whether current patent law is concerned with consent to participation in research or consent to patent filing. While the latter may be preferred to avoid redundancy, the reality is that both should be in play in any given situation. That we can contemplate their coexistence begs the question of the standard to which research practice should be held. Let us consider four scenarios:

Scenario 1: 'deception of individuals or research groups'; (i) no valid consent to taking of sample, (ii) no valid consent to patenting

This scenario is akin to the facts of the *Moore* case. In these circumstances of deception or bad faith, then there is no valid consent either to the research or to its commercialization. On a principled approach and in terms of *HOWARD FLOREY/Relaxin* this would axiomatically be an abuse of individuals sufficient to justify the denial or revocation of a patent.

Scenario 2: 'an informed individual or research group'; (i) consent to taking of the sample, (ii) refusal to permit patenting

In this scenario where the individual or group has been informed of the prospect of patenting, there is a clear expression and division of attitude as between consent to the research and rejection of its commercialization. This, arguably, would also count as an immoral invention because on a first principles approach the very reason why we seek consent would be thwarted if it were otherwise. That is, if consent is a means to respect individuals' wishes, direct disregard for those wishes is self-evidently disrespectful. As argued above, if patenting were to be allowed in such circumstances, the onus would be on those advocating such a position to justify the eventuality.

Scenario 3: 'current practice'; (i) valid consent to taking of sample, (ii) no view on patenting

This scenario is termed 'current practice' because consent procedures in many research projects do not, as yet, contain specific mention of the prospect of commercialization. This does, however, vary considerably across jurisdictions and practice is generally changing.[36] None the less, in cases where there is no specific consent provision in respect of patenting, then we must ask whether a subsequent patent application would be 'immoral' in terms of patent law? How are we to judge this? An answer may lie, once again, in congruence between regulatory and patent regimes. If the legal regulatory standard requires commercial interests and practices to be disclosed, then a failure to obtain consent to patenting may be fatal to a subsequent application. However, if the regulatory regime does not, as yet, mandate such disclosures, then it is arguable that the patent system should not go beyond that regime to impose a higher standard. To do so would be tantamount to pre-emptive standard-setting, which is problematic for a number of reasons. It would allow patent law to drive wider regulatory considerations and would tempt pre-emptive moral judgements that may then take on the mantle of precedent. A ruling or even a series of rulings from the EPO, for example, if taken to be effective throughout any number of signatory states named in the patent application, would undermine the principle of moral pluralism, which the institutions of Europe have repeatedly sworn to respect (van Overwalle 2002). The general tenor of this argument is that regulatory law not patent law should lead the way. This is recognized to some extent in the EC Biotechnology Directive itself, where it is stated in Recital 14 that:

> Whereas a patent for invention does not authorise the holder to implement that invention, but merely entitles him to prohibit third parties from exploiting it for industrial and commercial purposes; whereas, consequently, substantive patent law cannot serve to replace or render superfluous national, European or international law which may impose restrictions or prohibitions or which concerns the monitoring of research and of the use or commercialisation of its results, notably from the point of view of the requirements of public health, safety, environmental protection, animal welfare, the preservation of genetic diversity and compliance with certain ethical standards.

In such circumstances, it would arguably be insufficiently offensive on the current interpretation of patent morality provisions to prohibit patenting. Indeed, it is likely that a further quote from the *H2 Relaxin* case would be taken to represent the position: 'human tissue or other material, such as blood, bone, and so on, has been widely used for many years as a source for useful products, . . . many life-saving substances . . . are isolated in this way and many have been patented'.

But standards and expectations shift constantly. And, to the extent that regulatory standards for information disclosure are moving to require disclosure of commercial interests or of commercialization practices then, once again, this could serve as an indication that evidence of consent to patent filing should form part of the application process.

Scenario 4: 'no consent to research required'; but what of consent to patenting?

It was suggested earlier that it is possible to conceive of circumstances where consent to research is neither practicable nor required, and where justifications for proceeding with research are found outside the consent paradigm. But in such circumstances, should consent to patenting remain a requirement? It is suggested that this is an example, a fortiori, of insisting that the patent system does not unduly encroach on the realms of regulation. To require this provision – even if such consent were obtainable[37] – would add additional burdens that the regulatory regime itself does not contemplate. And, while this is not to suggest that individuals may continue to object to commercialization of, say, the use of their samples even when their consent to that use is not strictly required, it would be tantamount to the patent system assuming a primary regulatory role to insist on consent to patenting none the less. In such circumstances this would be tantamount to usurpation of function.

11 FURTHER PRACTICALITIES, OBJECTIONS AND LIMITATIONS

How would evidence of valid consent form part of the patent application verification process? How can the patent system check that suitable consent has been obtained? What if there is evidence that even one person has not appropriately consented out of the research group? Is the whole patent then invalid? In sum, would the practicalities not be simply too cumbersome to be workable?[38]

While these are indeed onerous considerations, the issue is largely one of procedural regularity, and in procedural terms a presumption in favour of patentability could operate if evidence were submitted along with the patent application that appropriately worded consent forms had been employed during the research process. Evidence would already exist from within the regulatory system as to whether the requisite signatures had been obtained. This, too, could be submitted. If, however, the presumption were to be challenged at either the grant or revocation stage, it would beg the question of what standard of care would be required. Must there be evidence that everyone consented? That is, would less than 100 per cent compliance threaten the patent? Not only would this be unrealistic but it would be higher than the standard required within ethical regulatory review. Rather, if questions were raised over the validity of consent in respect of any given patent, a range of thresholds could be used to judge whether the patent should stand or fall.[39] This range could include:

- Was *reasonable care* exercised in drafting the consent form to include mention of the possibility of patenting?
- Was *reasonable care* exercised in obtaining the consent?

- Was any error committed in *good faith*?
- Was the failure in the consent process due to *bad faith*?
- Have *ethical review guidelines* and other regulatory requirements been followed?
- Have concerns or challenges been raised (and been successful) within the *ethical review mechanism*?

These criteria reflect a mix of questions about legal compliance and moral conduct. They are mere suggestions of factors that could be taken into account in what is – at the end of the day – a value judgement on the propriety of granting a monopoly right over the outputs of ethically questionable research. This is therefore not to suggest that a breach of a regulatory rule would be determinative of the outcome in every case (although it would often be highly persuasive), nor is it to promote an absolutist view that the failure to obtain a single research subject's signature, or even those of a relatively small group, would be fatal to an application. The process would require to be more balanced; more nuanced. By the same token, if Advocate General Jacobs' view expressed above is correct, namely 'the requirement of consent to *all* potential uses of human material may be regarded as fundamental', then we should be moving to a position where every consent form informs a potential research participant of future commercial or patenting possibilities as a matter of course within the regulatory regime. Any evidence-gathering role, therefore, would already have been largely discharged on behalf of any patent office lest it be called upon at a future date to consider the morality of a given invention on the grounds of lack of informed consent.

A final cautionary point flows from this. If consent to patenting were to form a recognized part of patent law then the focus would doubtless shift to the consent process itself. The obligation would be to explain commercialization policy (or at least the prospect of patenting) at the initial stages of obtaining consent to research involvement. The practical outcome of this, however, reveals a limitation with the concept of consent itself outlined above, namely, that it merely represents a right of refusal. Thus, the only power of those who are happy to be involved in research but who are uncomfortable with its commercialization is to refuse to take part at all. In most cases it will simply be unrealistic that researchers would accept research participants who presented with a disjuncture of consents, that is, consent to research but not to patenting. They would most likely be rejected from the project by the researchers themselves.[40] The reality may be, then, after all of this, that those who truly wish to participate in research will have no real choice but to accept commercialization policies. This brings us to another limitation of the consent process, namely, the imbalance of power between researchers and participants. In sum, the irony is that the question of consent to patenting may never reach the patent office enquiry stage. Although this may be reassuring for the patent examiners of Europe and beyond, it is vitally important to recognize that this conclusion does not diminish the value of the instant discussion. This has sought to demonstrate that mechanisms *can* be developed within patent law to adhere to the fundamental principle of respect for persons that underpins most modern legal systems. It just so happens that the tool that is currently available – consent – has its own inherent limits. But this does not detract from the central messages of this piece which are (1) the need for better understanding of the relationships between patent systems and other social systems, (2) the need for congruence across systems, and most especially in respect of fundamentals such as respect for persons, human dignity and the right of integrity, and (3) the recognition of the natural and appropriate boundaries between those systems. In this sense, consent is a mere distraction.

NOTES

1. For an example of this see the UK Biobank project and the deliberations and recommendations of the Interim Advisory Group, www.ukbiobank.ac.uk/ethicsgov.php, 12 December 2005.
2. An example of this is found in Art. 5 of the Clinical Trials Directive (EC Directive 2001/20/EC 2001). But note how other safeguards are then employed.
3. An example of this is the English and Welsh provisions under Section 60 of the Health and Social Care Act 2001 under which the Secretary of State can designate certain uses of information to be in the public interest. Applications for use are scrutinized by a body called the Patient Information Advisory Group (PIAG) to ensure the appropriateness of granting access.
4. Medicines and Healthcare Products Regulatory Agency (2006), *Investigations into Adverse Incidents during Clinical Trials of TGN1412*, available at: http://www.mhra.gov.uk
5. See, for example, UNESCO (1997), *Universal Declaration on the Human Genome and Human Rights*, Art. 5(b): 'In all cases, the prior, free and informed consent of the person concerned [in treatment or research] shall be obtained'; the Council of Europe (1997), *Convention on Human Rights and Biomedicine*, Art. 5: 'An intervention in the health field may only be carried out after the person concerned has given free and informed consent to it'; and the UNESCO *Universal Declaration on Bioethics and Human Rights* (2005), Art. 6: 'Any preventive, diagnostic and therapeutic medical intervention is only to be carried out with the prior, free and informed consent of the person concerned, based on adequate information. The consent should, where appropriate, be express and may be withdrawn by the person concerned at any time and for any reason without disadvantage or prejudice'.
6. For a discussion of this in the UK context, see Mason and Laurie (2006), Chapter 9.
7. Ibid., for discussion.
8. On this last point see Mason and Laurie (2001).
9. A rare example of strong and active involvement of research subjects in research design and development is the UK ALSPAC (Avon Longitudinal Study of Parents and Children) project, www.alspac.bris.ac.uk, 1 November 2006.

10. While examples can be found of research projects that required re-consenting of participants, this, arguably, simply gives a further opportunity to refuse and to leave the project in light of further information. It does not provide an occasion to influence the research or its direction.
11. The seminal decision from the House of Lords is *Sidaway* v. *Board of Governors of the Bethlem Royal Hospital* (1985). For a good, if somewhat dated, comparative analysis see Geisen (1988). For more recent treatment, see Hervey and McHale (2004).
12. The waters can be muddied by the fact that patients are often recruited to be research subjects by their own clinicians who themselves become the researchers thereby blurring the boundaries of the respective relationships and the attendant legal rights and obligations.
13. Group of Advisers on the Ethical Implications of Biotechnology (1996), Para. 2.4. This was reiterated most recently in European Group on Ethics (2002).
14. For intriguing argument and analysis to the contrary, see Beyleveld (2000).
15. In French it reads: 'considérant que, si une invention porte sur une matière biologique d'origine humaine ou utilise une telle matière, dans le cadre du dépôt d'une demande de brevet, la personne sur laquelle le prélèvement est effectué doit avoir eu l'occasion d'exprimer son consentement éclairé et libre à celui-ci, conformément au droit national'; and in Spanish it reads, 'Considerando que, cuando se presente una solicitud de patente de una invención que tiene por objeto una materia biológica de origen humano o que utiliza una materia de este tipo, la persona a la que se hayan realizado las tomas deberá haber tenido ocasión de dar su consentimiento libremente y con la debida información sobre dichas tomas, de conformidad con el Derecho nacional'.
16. Opposition Division (EPO) (2002). For commentary see Laurie (2004).
17. The same provisions appear in Art. 6(2)(c) of the Biotechnology Direction.
18. This reads: 'The human body, at the various stages of its formation and development, and the simple discovery of one of its elements, including the sequence or partial sequence of a gene, cannot constitute patentable inventions' (equivalent to Art. 5(1) of the Biotech Directive).
19. Wisconsin Alumni Research Foundation, 13 July 2004, Examining Division (EPO).
20. See the Opinion of AG Jacobs in *Kingdom of the Netherlands* v. *Council of the European Union and the European Parliament* (2002), Para. 192.
21. For a discussion in the biotechnology context, see Dutfield (2003).
22. Consider, for example, the work of the Commission for Intellectual Property Rights (2005), and the recent policy directions of the World Intellectual Property Organization (2005).
23. See *Kingdom of the Netherlands* v. *Council of the European Union and the European Parliament* (2002), Para. 80 and associated text.
24. At least two member states – Denmark and Belgium – have considered making a 'consent to patent filing' provision a formal part of their law since the adoption of the Directive. Neither, however, has succeeded, and it is the author's understanding that there are no current plans in either jurisdiction to renew such efforts: Personal Communication with Dr Bart Claes, European Patent Office, June 2004.
25. This is better left to the systems theorists: see, for example, Teubner (1988).
26. See the Opinion of Advocate General Jacobs in *Kingdom of the Netherlands* v. *Council of the European Union and the European Parliament* (2002), Paras. 191–5.
27. Consider: The Charter of Fundamental Rights of the European Union, Art. 3(2): Right to the Integrity of the Person – 'In the fields of medicine and biology, the following must be respected in particular . . . the free and informed consent of the person concerned'. Indeed, Advocate General Jacobs accepted 'without doubt' that this was a fundamental right within the meaning imported by the obligation on the European Union to respect such rights; ibid., Para. 197. Furthermore, Recital 16 of the Biotech Directive states: 'Whereas patent law must be applied so as to respect the fundamental principles safeguarding the dignity and integrity of the person'.
28. See note 26 above, Para. 211.
29. This is discussed further below.

30. See note 26 above and associated text.
31. For example, one might claim that a prime motivator within the regulation system is public safety while within the patent system the prime motivator is economic advancement.
32. The ECJ judgement might be seen as a defence of the current position but its failure to acknowledge a relationship between the patent system and other social systems raises serious questions about the robustness of that defence.
33. See, *Plant Genetic Systems* (1995), Item 7: 'a particular subject-matter shall not automatically be regarded as complying with the requirement of Article 53(a) EPC merely because its exploitation is permitted in some or all of the contracting states. Thus, approval or disapproval of the exploitation by national law(s) or regulation(s) does not constitute per se a sufficient criterion for the purposes of examination under Article 53(a) EPC'. See also the discussion in Opposition Division (EPO) (2002).
34. For a discussion of recent developments, see Laurie (2004).
35. The Administrative Council of the European Patent Office amended its guidelines to reflect the contents of the EU Directive on 16 June 1999, www.european-patent-office.org/epo/ca/e/16_06_99_impl_e.htm, 12 December 2005.
36. See, for example, the guidance from the Medical Research Council (2001), Section 4, and the opinion of the Group of Advisors on the Ethical Implications of Biotechnology to the European Commission 1996, Para. 2.4.
37. Often research is done on an anonymous basis in order to protect privacy. This privacy protection might be compromised by attempts to obtain consent.
38. AG Jacobs certainly thought so in the Dutch challenge to the Biotech Directive, note 25 above, Para. 212: 'to make evidence of . . . consent a condition of granting a biotechnological patent – presumably by way of the morality principle – to my mind risks being unworkable. Biotechnological inventions may be derived from research on possibly thousands of blood or tissue samples, possibly pooled and almost certainly anonymized at the time of analysis. I do not consider it reasonable to expect patent examiners to satisfy themselves that the chain of consent with regard to each sample is unbroken and evidenced'.
39. An additional complicating factor would be the need to demonstrate a causal connection between a particular research protocol for which consent had been obtained and the eventual downstream invention that is the subject of the particular patent application. That is, does the particular invention have a sufficiently strong connection to the original research done with research subjects to require the latter's consent to patenting?
40. It is also highly likely that consent to patenting could not be withdrawn once obtained for obvious reasons.

REFERENCES

Beyleveld, D. (2000), 'Why Recital 26 of the EC Directive on the Legal Protection of Biotechnological Inventions Should Be Implemented in National Law', *Intellectual Property Quarterly*, **4**, 1–26.

Beyleveld, Deryck and Roger Brownsword (2001), *Human Dignity in Bioethics and Biolaw*, Oxford, UK: Oxford University Press.

Chester v Afshar [2004], 3 WLR 927, HL.

Commission for Intellectual Property Rights (2005), www.iprcommission.org, 12 December 2005.

Council of Europe (1997), *Convention on Human Rights and Biomedicine*, Art. 5, Strasbourg, France: Council of Europe, http://conventions.coe.int/Treaty/EN/Treaties/Html/164.htm

Cragg Ross Dawson (2000), *Public Perceptions of the Collection of Human Biological Samples*, London, UK: Wellcome Trust/Medical Research Council, www.wellcome.ac.uk, 12 December 2005.

Dutfield, Graham (2003), *Intellectual Property Rights and the Life Science Industries: A 20th Century History*, Aldershot, UK: Ashgate Publishing.
EC Directive 98/44/EC (The Biotechnology Directive) (1998), On the Legal Protection of Biotechnological Inventions, *Official Journal* L213/13.
EC Directive 2001/20/EC (2001), On the Approximation of the Laws, Regulations and Administrative Provisions of the Member States relating to the Implementation of Good Clinical Practice in the Conduct of Clinical Trials on Medicinal Products for Human Use, *Official Journal* L300/58.
European Group on Ethics (2002), *Ethical Aspects of Patenting Inventions Involving Human Stem Cells*, Opinion No. 16, Brussels, Belgium: European Commission.
Geisen, Deiter (1988), *International Medical Malpractice Law*, London, UK and Boston, USA: Martinus Nijhoff.
Group of Advisors on the Ethical Implications of Biotechnology (1996), *Ethical Aspects of Patenting Inventions Involving Elements of Human Origin*, Opinion No. 8, Brussels, Belgium: European Commission.
Hapgood, Rhydian, Chris McCabe and Darren Shickle (2004), *Public Preferences for Participation in a Large DNA Cohort Study: A Discrete Choice Experiment*, Sheffield, UK: Sheffield Health Economics Group, www.shef.ac.uk/~shef/discussion/04_5FT.pdf, 12 December 2005.
Hervey, Tamara K. and Jean V. McHale (2004), *Health Law and the European Union*, Cambridge, UK: Cambridge University Press.
HOWARD FLOREY/Relaxin [1995], EPOR 541.
Kingdom of the Netherlands v Council of the European Union and the European Parliament (2002), C-377/98, All ER 97.
Laurie, G. (2004), 'Patenting Stem Cells of Human Origin', *European Intellectual Property Review*, **26**(2), 59–66.
Llewelyn, M., D. Beyleveld and R. Brownsword (2000), 'The Morality Clause of the Directive on the Legal Protection of Bio-technological Inventions: Conflict, Compromise and the Patent Community', *Pharmaceutical Biotechnology and European Law*, 157–81.
Mason, K. and G. Laurie (2001), 'Consent or Property? Dealing with the Body and its Parts in the Shadow of Bristol and Alder Hey', *Modern Law Review*, **64**(5), 710–29.
Mason, Kenyon and Graeme Laurie (2006), *Mason and McCall-Smith's Law and Medical Ethics*, Oxford, UK: Oxford University Press.
Medical Research Council (2001), *Human Tissue and Biological Samples for Use in Research*, London, UK: Medical Research Council.
Moore v Regents of the University of California (1990), 793 P. 2d 479 (Cal. 1990).
O'Neill, Onora (2002), *Autonomy and Trust in Bioethics*, Cambridge, UK: Cambridge University Press.
Opposition Division (EPO) (2002), 'The Edinburgh Patent', 21 July 2002, Para. 2.5.3.
Pearce v United Bristol Healthcare NHS Trust (1998), 48 BMLR 118.
Plant Genetic Systems [1995], T 356/93, EPOR 357.
Sidaway v Board of Governors of the Bethlem Royal Hospital [1985], 1 AC 871.
Teubner, Gunther (ed.) (1988), *Autopoietic Law: A New Approach to Law and Society*, Berlin, Germany and New York, USA: Walter de Gruyter.
UNESCO (1997), *Universal Declaration on the Human Genome and Human Rights*, Paris, France: UNESCO, http://portal.unesco.org/shs/en/ev.php-URL_ID=1881&URL_DO=DO_TOPIC&URL_SECTION=201.html

UNESCO *Universal Declaration on Bioethics and Human Rights* (2005), Art. 6, Paris, France: UNESCO, http://portal.unesco.org/shs/en/ev.php-URL_ID=1372&URL_DO=DO_TOPIC&URL_SECTION=201.html

van Overwalle, Geertrui (2002), *Study on the Patenting of Inventions related to Human Stem Cell Research*, Luxembourg: European Commission, pp. 68–9, http://ec.europa.eu./european_group_ethics/publications/index_en.htm, 12 December 2005.

World Intellectual Property Organization (2005), www.wipo.int, 12 December 2005.

12. Reshaping bio-patents: measures to restore trust in the patent system

Geertrui Van Overwalle

1 INTRODUCTION

Recent events have convincingly shown that public opinion is very sensitive to possible unfair use of the patent system in the field of genetic inventions. Witness the strong reactions evoked by the grant of three patents covering the breast cancer gene, its mutations and the diagnostic and therapeutic applications based on the gene's sequence by the European Patent Office (EPO) and by the restrictive licence policy of the patent holder Myriad Genetics.

The first part of this chapter will analyse the objections that have been put forward against the patenting of biological material over the years. Initially, objections to deny biological material patent protection mainly related to the specific nature of living organisms, the lack of novelty, inventive step or industrial applicability, the impossibility of description and the non-reproducibility of the method of making. However, in recent years, new objections have been raised against bio-patents and a clash between patents and social and ethical concerns appears to have materialized. The second part will examine measures and tools that might help to remove the public reserve with regard to the current use of bio-patents. Instruments that will be examined are the toughening of patent standards, the implementation of an origin and informed consent requirement, a renewed look at the scope of protection, the monitoring of patent proprietors' licensing practices and the creation of patent pools and compulsory licences for public health. Such tools, either internal or external to patent law, should be used in order to reduce the negative social impact and to restore trust in the patent system.

2 OBJECTIONS AGAINST BIO-PATENTS

Close reading of academic writers and court decisions reveals that a series of objections is put forward to deny biological material patent protection (pre-grant objections) and that a chain of concerns is raised to point to potential negative effects of patent rights (post-grant concerns). Pre-grant objections relate to patentable subject matter, patentability requirements and disclosure. Post-grant concerns mainly relate to the blocking effects of patents on research and health care (see Figure 12.1).

2.1 Subject Matter

A major objection that has often been put forward against patents for biological material is that biological material is not the result of a creative process. Micro-organisms, plants and animals are products of nature and are non-inventions or 'Nicht-Erfindungen' lacking 'Erfindungsqualität' (Delcorde 1985; Mathély 1978; Moufang 1991; Mousseron 1984; Paterson 1992; Van Overwalle 1996; Van Overwalle 1999). Consequently, they do not qualify as patentable subject matter.

In 1969 this objection was firmly refuted by the German Bundesgerichtshof in the *Rote Taube* (Red Pigeon) case by arguing that a technical invention can also exist in the systematic application of biological forces of nature: 'Als patentierbar kann werden angesehen eine gewerblich verwertbare neue fortschrittliche und erfinderische Lehre zum planmässigen Handel unter Einsatz beherrschbarer Naturkräfte zur Erreichung eines kausal übersehbaren Erfolgs'[1] (An industrial applicable, novel, advanced and inventive technical teaching, using controllable forces of nature in order to achieve a causal surveyable outcome, is considered to be patentable) (Van Overwalle 1997; Van Overwalle 1999, pp. 143–94). In the US, the Supreme Court followed the same line of reasoning in *Diamond v Chakrabarty* in 1980.[2]

In 1998 the issue was settled in Europe through an express provision on the patentability of biological material in an EU Directive on the legal protection of biotechnological inventions (EC Directive 98/44/EC 1998). The Directive stipulates that 'inventions which are new, which involve an inventive step and which are susceptible of industrial application shall be patentable even if they concern a product consisting of or containing biological material or a process by means of which biological material is produced, processed or used' (Article 3.1) and that 'biological material which is isolated from its natural environment or produced by means of a technical process may be the subject of an invention even if it previously occurred in nature' (Article 3.2).

Notwithstanding the explicit safeguard offered in the Directive, the product of nature objection persists, even with more vigour, against DNA sequences. DNA sequences are said to be discoveries, rather than inventions and thus unfit for patent protection (Knoppers 1999). Patent offices in Europe and the US, however, developed the view that genes are to be

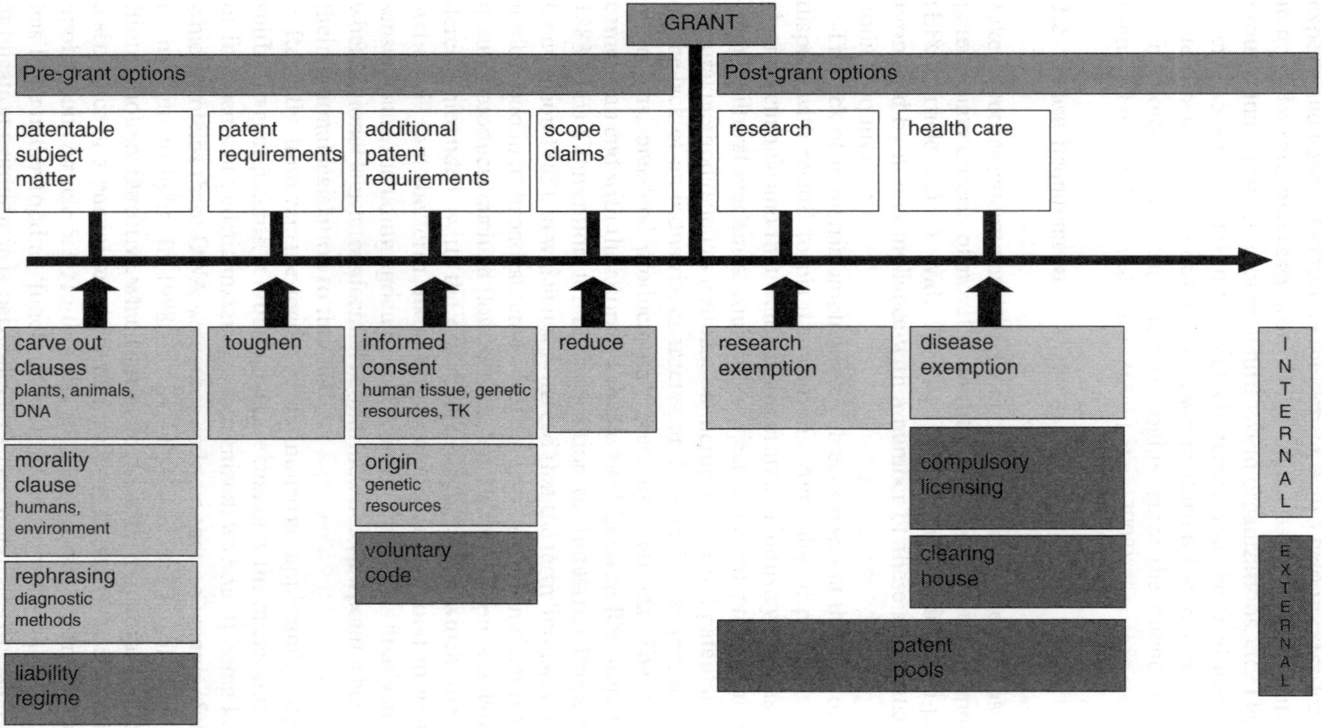

Figure 12.1 *Overview of the concerns (above the arrow) and remedies, both internal and external to patent law (below the arrow), with regard to bio-patents*

treated as chemical substances and are patentable, if human technical intervention is required to identify and reproduce them. This approach gained wide acceptance in legal and patent circles (Andrews 2002; Ducor 1998; Straus 1995). An express provision on the patentability of genes is officially promulgated in Europe by way of the EU Biotechnology Directive. The key that the Directive offers lies in the distinction between genes 'as they occur in nature', which are not patentable, and genes 'that are isolated from their natural environment by means of a technical process', which are patentable. In this regard, the Directive puts forward that the human body, at the various stages of its formation and development, as well as the simple discovery of one of its elements, including the sequence or partial sequence of a gene, cannot constitute patentable inventions (Article 5.1), whereas an element isolated from the human body or otherwise produced by means of a technical process, including the sequence or partial sequence of a gene, may constitute a patentable invention, even if the structure of that element is identical to that of a natural element (Article 5.2). The Directive, including Article 5, was implemented in the Implementing Regulations of the European Patent Convention (EPC, 1973) in 1999,[3] thus providing a legal ground for the European Patent Office (EPO) patent practice in the field of human genomics.

However, the controversy lingers on and many critics do not acknowledge the distinction set forth in the EU Directive on biotechnological inventions (Beyleveld and Brownsword 1997; Jacobs and Van Overwalle 2001). For them, genes – whether they are isolated from their natural environment or not, with knowledge of their function or otherwise – are, in principle, not patentable for ethical, philosophical or religious reasons. And they hold the view that performing some 'technical intervention' does not legitimize gene patents.

The current debate surrounding DNA patents reveals a dual development. First, there is a shift in the character of the objections. Over a very long period the living matter objection raised against patent protection for biological material was of a technical nature: biological matter did not meet the standard of invention concept. Recently, however, the living matter objection has been ethically qualified and biological material should not be patented for ethical reasons: the human genome is viewed as the common heritage of mankind and thus considered unfit for any kind of appropriation. The ambivalence of the new technology has rightly made the necessity of ethical monitoring even more urgent. The confrontation of ethics and patent law is difficult, to say the least, and usually ethics are seen in this context as a disturbance, as the *grosse Störung* (big disruption) (Van Overwalle 1996). Second, there is a shift in the forum where the discussion is being held. Initially, the discussion largely took place in legal circles of experts and courts. Nowadays the debate is also held in extra-legal circles: in non-governmental organizations, consumer associations, environmental groups, animal rights' groups, Third World organizations, etc. The opinion of civil society is vital in the biotech debate and the final acceptance of biotech patents. Some observers, however, warn for increased public interest in biotech intellectual property rights, since the public is too often premised on fundamental errors (Gold et al. 2002).

2.2 Patent Requirements

Like any other patent, patents relating to biological material must meet the patentability criteria of novelty, inventive step and industrial applicability (EPC Article 52). A wide series of objections against bio-patents is prompted by non-compliance with a number of these substantial patentability requirements.

The lack of industrial applicability objection was at the core of a heated dispute with regard to plant inventions. Agricultural products that were clearly definable and that were manufactured in industry, such as fertilizers or agricultural machines, could be subject to patent protection, but agricultural methods such as fertilization methods were not patentable because of the lack of an industrial character, and the same was true for breeders' inventions, breeders' products and breeding methods. The controversy came to an end with the adoption of the 1934 London Revision Act of the 1883 Paris Convention for the Protection of Industrial Property (Paris Convention 1883), in which it was agreed that the term 'industry' should be understood in its broadest sense, including not only manufactured but also natural products such as flowers. Besides, this objection was largely rendered unfounded, with the enactment of the EPC, which stipulated in Article 57 that the term 'industry' should be understood in its broadest sense, that is, including agriculture. The impediment was finally invalidated when national legislators formally confirmed this viewpoint when adapting their national legislation to the EPC.

Recently, however, the focus of the industrial applicability debate has shifted to gene patents. It is often the case that only the nucleotide sequence of fragments of genetic material is recognized, without it being known for which proteins these DNA sequences code, so that no meaningful use can be brought to light. In 1998 this problem was largely solved in the EU Biotechnology Directive, which stipulates that the industrial application of a sequence or a partial sequence of a gene must be disclosed in the patent application (Article 5.3). In this regard the notion 'industrial application' has to be understood as 'function': in order to comply with the industrial application criterion it is necessary in cases where a sequence or partial sequence of a gene is used to produce a protein or part of a protein, to specify which protein or part of a protein is produced or what function it performs (Recital 24).

The lack of novelty and inventive step was mainly raised in the past against patent protection with regard to plant inventions: it was repeatedly argued that breeders' products missed novelty and that plant varieties bred by traditional methods were not beyond the grasp of the ordinary artisan (Van Overwalle 1996; Van Overwalle 1999). However, the molecular developments in (plant) biotechnology over the past years made it possible to overcome the impediments of novelty and inventiveness in lots of cases. The lack of novelty and inventive step is at present also being voiced against DNA patents, especially genomic DNA patents. As biotechnological techniques are becoming more and more established, it could very well be that a specific gene transfer will increasingly have to be more difficult or uncommon to meet the requirement of inventiveness of patent law.

2.3 Scope and Disclosure

Like any other patent, bio-patents must disclose the invention in a manner sufficiently clear and complete for it to be carried out by a person skilled in the art (EPC Article 83). In the past this disclosure requirement equally aroused emotions, in particular the impossibility of describing both the method of making and the end product, as well as the unfeasibility of repeating the original process of making (Van Overwalle 1996; Van Overwalle 1999). Recent studies suggest that many existing bio-patents still suffer from a lack of enabling disclosure (Bostyn 2001; Nuffield Council on Bioethics 2002).

Closely related to the disclosure requirement is the issue of claim drafting and scope of protection. Over the past years, the scope of biotech patent claims has increasingly been put into question: claims are unduly broad in comparison with the disclosed and described invention. Broad protection might lead to an illegitimate over-reward for the patentee.

2.4 Effects of Patents

Lately, a number of voices have been heard, expressing great concern with regard to the effects of patents. Some studies have pointed at the possible negative effects of bio-patents on scientific research in general and genomic patents in particular (Lecrubier 2002; Nelkin 2002; Nuffield Council on Bioethics 2002; Wolfrum and Stoll 2002). The effect of patents is somehow tempered by the research exemption and it is generally accepted that in an academic setting the use of patented gene technology is admissible, provided the research is not intended for commercial use. However, recent studies on drug patents have shown that the exact scope of this exemption differs greatly from country to country, because different national legislations and court rulings exist on this matter (Van Overwalle 2000a; Van Overwalle 2000b).

Some surveys suggest that patents also might have possible detrimental effects on the development and validation of clinical tests (Merz et al. 2002), medical progress in general (Bobrow and Thomas 2001) and on diagnostics and health care in particular (Lecrubier 2002; Nelkin 2002; Sevilla et al. 2002). Blocking effects might occur, in two different ways. First of all, access to and use of the patented technology might be hampered by the unreasonable licensing policy of a patent owner. In this regard, the licensing approach as put into practice by Myriad Genetics for the screening of the BRCA1 and BRCA2 genes has given rise to a strong and worldwide reaction. Second, access and use might be hindered by the existence of a 'patent thicket': a dense web of overlapping patents a researcher or a company must hack its way through in order to actually develop and commercialize a new product (Scherer 2002; Shapiro 2001). The patent thicket might in its turn lead to 'royalty stacking': a heap of multiple licence fees, which a downstream inventor has to pay to upstream patent holders (Merz et al. 2002; Nicol and Nielsen 2003; Straus et al. 2002; Walsh et al. 2003), which deter researchers and companies from entering the field.

3 MEASURES TO MEET CONCERNS

The many and varied objections and concerns raised with regard to biotech patents should be taken to heart. The hesitant or dismissive attitude of many people with regard to biotech-patents should not be trifled with. Seriously taking into account these concerns is the only approach that will be beneficial to industry in the long run. Only if the public accepts them, will biotechnological products continue to find their way on to the market. Ignoring the public reserve with regard to the current use of patents by the bio-industry might turn out to be injudicious. One day the bio-industry might find itself being confronted with a public opinion taking the view that ethical concerns and moral principles are more important than greater competitiveness, economic growth and biotechnological development (Gold 2000; Van Overwalle 2005a).

It is therefore worthwhile and timely to consider the measures that patent law offers for addressing some of the objections and concerns arising out of biotechnology patenting.

3.1 Subject Matter

One way to deal with ethical and social concerns with respect to the patenting of plants and animals is to legislate a carve-out from patent protection. The legal basis to do so is the Agreement on Trade-related Aspects of Intellectual Property Rights (TRIPs Agreement 1994), which came about as an Annex Treaty to the WTO Agreement and establishes the key international standard with regard to IPR protection.[4] In the field of life sciences, biotechnology and genetic engineering the TRIPs Agreement contains a provision that is highly relevant in this respect. Article 27.3 stipulates that member states *may* exclude plants and animals – not micro-organisms – from patentability. Various studies point to the flexibility offered in this provision to reassess patentable subject matter to meet the needs of developing countries (Commission on Intellectual Property Rights 2002). With regard to developed countries, however, it is probably futile to discuss this possibility, because there are already so many precedents (Murashige 2002).

Another way that has been put forward to deal with the reluctant attitude of patents on DNA is to lock out DNA sequences from patent law. It has been suggested that patents should no longer be granted for DNA, but only for new medicinal products, new vaccines and genetic tests that are developed on the basis of DNA, for example, drugs to combat cardiovascular diseases, cancer, infectious diseases, diabetes or fertility problems, new vaccines against HIV or flu and genetic tests for disorders such as Huntingdon's disease or cystic fibrosis. In other words, only patents on the end products would be granted, while patents on the base product used to manufacture these products would be refused. In technical terms, this means that there would be no more patents on DNA as a research tool (product claims), but there would be patents on the use of DNA to diagnose, prevent or treat a specific disease (use claims) and on the resulting end product. Patents on products that monopolize 'life' or the genetic raw materials of life would thus be avoided and patents would only be granted on the resulting products (Bobrow and Thomas 2001; Jacobs and Van Overwalle 2001; Straus 2003). In this line of this reasoning, Germany announced it would implement the Directive, but at the same time initiate an amendment process on the European level to support the improvements and clarification needed, in particular for examining the scope of product patents in the biotech field.[5]

A third way to deal with ethical concerns relating to specific categories of human (and animal) material has been established through the insertion of the morality clause in the EU Biotechnology Directive. The Directive explicitly stipulates that inventions, the commercial exploitation of which is considered to run counter to *ordre public* or morality, shall be excluded from patentability (Article 6.1). In an effort to offer some concrete guidance for the interpretation of the twin concept of *ordre public* and morality, the Directive has taken up a non-exhaustive list of inventions that shall be considered unpatentable: processes for cloning human beings, processes for modifying the germ line genetic identity of human beings, uses of human embryos for industrial or commercial purposes and processes for modifying the genetic identity of animals, which are likely to cause them suffering without any substantial medical benefit to man or animal, and also animals resulting from such processes (Article 6.2). This list is an illustrative list of inventions excluded from patentability and, obviously, it is not intended to be exhaustive (Recital 38). The scope of this Directive and its application to human stem cells appears to be uncertain. The current controversy on the patentability of stem cells will probably lead to the exclusion of embryonic stem cells from patent law (Van Overwalle 2002; Van Overwalle 2004a; Van Overwalle 2005b).

In relation to the list of exclusions, the more fundamental question arises whether the list should be taken up in patent law after all. There is a strong feeling that Article 6.2 does not aim at limiting the patent implications of certain biotech inventions, but wishes to exclude certain fields of research as such. We agree that patent law should take into account ethical concerns and that patent law can and should act as a 'moral tollbooth' (Gold and Caulfield 2002), but patent law should only do so to the extent that it concerns matters that are directly and inextricably linked to *patents* and the exercise of patents; patent law should not interfere when *research* is ethically undesirable. Since a direct link is missing between ethics and patents in Article 6.2, we take the view that this provision should be abolished and that the exclusions should be treated in research regulations (Van Overwalle 2004a).

The TRIPs-equivalent of the morality clause is even wider in scope and might open the door for excluding additional subject matter from patentability, in particular inventions, of which the prevention within their territory of the commercial exploitation is necessary to protect 'plant life or health or to avoid serious prejudice to the environment' (Article 27.2). Some Member States have implemented the wide TRIPs wording in their national patent legislation,[6] but no information is as yet available as to the use that has been made of this exclusionary provision.

A fourth measure for restricting patentable subject matter in the field of biotechnology and for meeting the concerns of civil society, is to exclude genetic methods of testing from patent law. As could be read from a decision from the technical board of appeal at the EPO, this can be achieved by interpreting the scope of the current exclusion for diagnostic testing laid down in Article 52.4 EPC, in the sense that not only diagnostic methods 'practised on the human or animal body' (in vivo) are excluded, but also separate steps that are part of diagnostic methods and practised outside the body (ex vivo).[7]

Last but not least, a fifth possibility comes into sight. Eminent legal scholars have reflected on alternative regimes that enable entrepreneurs to appropriate the fruits of their investments in cumulative and sequential innovation without impeding follow-on innovation and without creating barriers to entry, in order to open new ways for stimulating investment in subpatentable innovation without impoverishing the public domain. One system that meets these concerns is a compensatory liability regime (Calabresi and Melamed 1972; Reichman 2000). Further research and experience should prove to which degree the proposed system is feasible and useful.

3.2 Patent Requirements

A strategy to meet the problems perceived with regard to patent requirements is to make patents more difficult to obtain, by raising the bar of the patent standards of novelty, inventive step and industrial applicability. Patent requirements as such are framed within the statutory context, but they are shaped in large part by the policies of the patent offices. Patent offices can strengthen the patent standards in their daily granting practice. Toughening patent biotechnology standards has already led to some significant changes in the field of DNA patenting.

However, thoughtful observers point to some critical issues in this regard. The development of higher patent standards might take some time, since it is only with the accumulation of experience that patent offices and courts can adapt existing law to reflect the realities of biotechnology. The ever-changing nature of biotechnology might have as an effect that by the time courts and patent offices have adapted their standards to a particular technology, the technology has become outdated (Gold 2002). Patent standards are not always free from interpretation. Of the three criteria for patentability, only novelty is relatively free of interpretation: either something is part of the state of the art, or it is not (Murashige 2002). Inventiveness is far from an objective standard as a practical matter and industrial applicability (Europe) and utility (US) can be equally problematic, especially with regard to genomic sequences.

Another attempt to give wider echo to ethical issues in a patent context is the introduction of additional patent requirements. The EU Biotechnology Directive (EC Directive 98/44 1998) offers a basis for the introduction of both an informed consent requirement and an origin requirement.

The EU Biotechnology Directive tries to give something to hold on to people, who in the framework of a diagnosis or treatment, give body material, which material is subsequently being used for scientific research (further use) and which forms the basis of a new drug or therapy, for which patent protection is requested and granted. Recital 26 stipulates in this context that 'Whereas if an invention is based on biological material of human origin or if it uses such material, where a patent application is filed, the person from whose body the material is taken must have had an opportunity of expressing free and informed consent thereto, in accordance with national law'. Recital 26 attempts to embed the general principle of free and informed consent of the donor of body material as proclaimed in Article 22 of the Convention on Human Rights and Biomedicine of 1997, in the context of biotech-patenting.[8] Although the discussion on the legitimacy of inserting a recital as a full provision in patent law has not ceased amongst legal scholars, various EU Member States are considering introducing an informed consent requirement in their patent laws (Van Overwalle 2003; Van Overwalle 2004b; Van Overwalle 2005a).

The EU Biotechnology Directive also shares public concern surrounding bioprospecting. The search for new pharmaceutical, biotechnological or agricultural applications has led to a growing interest in genetic resources and indigenous traditional knowledge. The contribution of indigenous communities has tended, however, not to be rewarded when the resulting products have been patented and commercialized. In order to remedy this problem, various countries turned to the Convention on Biological Diversity of 5 June 1992 (CBD 1992) for some guidance. Especially the third CBD objective, the so-called equitable sharing objective embodied in Articles 1, 8(j) and 15 of the CBD, triggered many signatory states to reflect upon appropriate legal instruments. Attention has been given to measures taken by provider countries to implement adequate measures (Van Overwalle 2005c). A very popular tool in this respect, which is now gaining worldwide interest, is a *sui generis* right termed either 'Collective Community Intellectual Property Right (CCIPR)' (Leistner 2003) or 'Traditional Intellectual Property Right (TIP right)' (Cottier 1998; Cottier and Panizzon 2004).

Attention has also focussed on the promotion of a range of measures that user countries, particularly developed countries, could implement in their patent law to put the CBD objectives to work (Van Overwalle 2005c). In this attempt, the EU Biotechnology Directive introduced an origin indication condition in patent law. Recital 27 of the Preamble of the Directive stipulates that 'Whereas if an invention is based on biological material of plant or animal origin or if it uses such material, the patent application should, where appropriate, include information on the geographical origin of such material, if known; whereas this is without prejudice to the processing of patent applications or the validity of rights arising from granted patents'. Although the debate amongst scholars continues as to the nature of such a requirement – should it be a substantial or a formal requirement? Should it be a patent requirement or a requirement outside patent law? Various EU Member States have started implementation of this requirement in their patent legislation (Van Overwalle 2004a; Van Overwalle 2005c).

A third way to take into account ethical concerns regarding the patenting of inventions based on biological material is by setting up mechanisms outside (patent) law, more in particular by establishing voluntary codes of conduct. A voluntary code of conduct is a standard that sets forth principles to guide a company's performance. Although a 'standard' sets forth rules or guidelines, compliance is not mandatory. One of the aims of a code often is to protect the public image of a company; after all, large corporations often aim at convincing consumers that they act morally. Hence, the penalty for nonconformity with a standard comes from the marketplace or the public and the threat of bad publicity (Echols 2003). These days, the issue of informed consent and geographical origin has become a matter of great public concern. Yet, if the (national) government's hands are tied or political consensus is hard to achieve, private sector standards are an option. If it appears undesirable to implement the informed consent or origin requirements through legal action, the establishment through voluntary codes of conduct might be considered (Van Overwalle 2002).

3.3 Scope

Yet another tool to address ethical and social concerns is the determination of the size or scope of patents on materials and methods in biotechnology. Patent offices and courts may reduce the scope of patent claims that are excessively broad in order to comply with legal requirements. In doing so, offices and courts are challenged to strike the correct balance between a fair protection for the patentee and a reasonable degree of access (Gold 2002; Resnik 2003).

3.4 Effects of Patents

As to remedying some perceived negative effects of biotechnological patents on research and health care, there are a few possible options. A first possibility is to exempt certain types of activity from infringement. A common exemption in this regard is the research exception. In the US the exemption is not part of the patent act but does exist as a judicially created doctrine. However, the doctrine is very narrow, and, as it currently stands, the exemption only applies to research undertaken for purely philosophical reasons or academic purposes with no prospect of commercialization whatsoever. The experimental use has to be limited 'solely for amusement, to satisfy curiosity or for strictly philosophical inquiry' in order to enjoy the exception.[9] Voices are raised expressing concern about this extremely limited interpretation (Saunders 2003; Sung and Maisano 2003). In Europe, the research exemption is part of patent law, but the exact scope of this exemption differs greatly from country to country, for different national legislations and court rulings exist on this matter (Chrocziel 1984; Cornish 1998; Straus 1993; Van Overwalle 2000d). The situation, both in the US and Europe, can be resolved by reinforcing and clarifying the research exemption, and by carefully defining the delicate border line between commercial and non-commercial research in the field of biotechnology and biomedicine.

Another exception to the rights of the patentee is the health care provider exemption. In the US the patent act exempts medical personnel and institutions from infringement if they are merely carrying out a patented method of treatment that does not involve a patented or regulated drug or device (Murashige 2002). In Europe a similar result has been achieved by denying patent protection for methods for treatment and diagnostic methods practised *on* the body (Article 52.4 EPC). In an attempt to meet concerns about genetic testing, diagnostic methods practised ex vivo might be denied patent protection (see above) or – in case this is not successful – the introduction of a wide disease exemption might be advocated.

A third possibility to meet ethical and economic concerns is to resolve problematic licensing practices. A few scenarios can be put to work here. When access to and use of the patented technology is hampered by a paramount patent, a compulsory licence might bring relief. In such cases the patentee is compelled by the government to license rights at reasonable rates. Although it does not contain the term 'compulsory licensing', the TRIPs Agreement affirms the right of Member States to grant compulsory licences and confirms their autonomy to determine the grounds upon which such licences can be granted (Article 31). In November 2001 the Ministerial Conference of the World Trade Organization (WTO) in Doha, Qatar even formally confirmed this point of view and stated that TRIPs should be interpreted in a manner supportive of public health. The formally recognized relationship between TRIPs and public health is extensively discussed by expert commentators with regard to developing countries (Abbott 2002; Vandoren 2002), but has hardly been explored in a developed country setting, as a tool to temper the consequences of an extreme monopoly. However, the use of compulsory licensing to resolve health concerns is gaining wide interest in developed countries as well.

Some countries have even designed, apart from the general compulsory licence for non-working, a specific public health licence. France has recently implemented an ex officio licence for national public health reasons in its patent act[10] and Belgium has proposed a special compulsory licence regime for national health reasons ('*dwanglicentie in het belang van de volksgezondheid*') (Van Overwalle 2004b). Unlike some European countries, the US has no compulsory licensing provision in its patent laws (Resnik 2003), let alone a specific compulsory licence to meet health concerns. Alternatively, it would be possible to legislate a change from current law.

When access and use are hindered by the existence of multiple patents, held by multiple patent owners, various mechanisms might help to clear the so-called patent thicket. Patent pools have been suggested as an alternative solution for the emergence of patent thickets. A patent pool is an agreement between two or more patent owners to license one or more of their patents as a package to one another or to third parties willing to pay the royalties associated, either directly by patentees to licensees or, indirectly, through a new entity specifically set up for administering the pool (Clark 2000; Klein 1997; Merges 2001). Clearing house models might also be a way to facilitate access, through a statutory licensing scheme. Such a patent licence scheme is administered by a collecting society, comparable in structure and function to the existing copyright societies for playing music on air (Gold 2002; Krattiger 2004). Similarly, users of genetic inventions would pay an equitable fee to the society for using the invention patented. The primary difference between a patent pool scheme and a clearing house mechanism is that a patent pool is a contract between multiple parties, whereas a clearing house is a statutory framework implemented by the government: patent pools provide a regularized transactional mechanism for substituting the statutory property rule baseline, which requires an individual bargain for each transaction.

4 CONCLUSIONS

Over the last years the concerns surrounding biotech patents have increased dramatically. The many and varied objections raised with regard to biotech patents should be taken to heart. The hesitant or dismissive attitude of many people with regard to biotech patents should not be trifled with. The answer to those concerns, however, will probably not be the carving out of biological material from patent law. There are other measures available that reduce the negative impact of biotech patents while preserving their benefits. Any policy framework for regulating and controlling biotechnology patents should incorporate parts of these proposed solutions. Only then will it be possible to take away public reserve against patents and restore trust in the benefits of the patent system.

NOTES

1. *Gewerblicher Rechtsschutz und Urheberrecht* (1969), 672 (with an annotation from Heydt); *International Review of Industrial Property and Copyright Law* (1970), 136.
2. Supreme Court, June 16 1980, 447 US 303; 65 L.Ed.2d 144; 100 S.Ct 2207, 206 USPQ 193. See also Van Overwalle 1997.
3. The Directive was inserted as Rule 23 (e) (2) EPC.
4. The multilateral trade negotiations in the GATT Uruguay Round, which were concluded in 1993 and resulted in the formation of the World Trade Organization (WTO), encompassed for the first time discussions on aspects of intellectual property rights. The result of those negotiations was embodied in the 'Agreement on Trade-related Aspects of Intellectual Property Rights' (TRIPs Agreement), contained in an Annex to the WTO Agreement. Membership of the WTO implies adherence to the TRIPs Agreement.
5. The rumour is spread that the German Justice Department is considering introducing a new law in Germany, which would prohibit *all* product per se protection in Germany and that the German Government might try to have a similar amendment made to the EPC. It has also been reported that EPO staff members have made personal statements in the media, questioning the absolute product protection as provided in the EPC. Also notice the European Parliament Resolution of 4 October 2001 on the Patenting of BRCA1 and BRCA2 ('Breast Cancer') Genes (B5-0633, 0641, 0651 and 0663/2001) calling on the Council, the Commission and the Member States 'to adopt the measures required to ensure that the human genetic code is freely available for research throughout the world and that medical applications of certain human genes are not impeded by means of monopolies based on patents' (Document B5-0633, 0641, 0651 and 0663/2001, Point 4).
6. For example, Belgium, see Van Overwalle 2000c.
7. Referral by the President of the EPO to the Enlarged Board of Appeal based on two divergent decisions of the Boards of Appeal (T 385/86, OJ EPO 1988, 308 ff. – *Bruker* and others and T 964/99, OJ EPO 2002, 4 ff. – *Cygnus, Inc.* and others), December 19 2003.
8. Art. 22 stipulates: 'When in the course of an intervention any part of a human body is removed, it may be stored and used for a purpose other than that for which it was removed, only if this is done in conformity with appropriate information and consent procedures'.
9. *Madey* v. *Duke University*, 307 F. 3d 1351 (Fed. Cir. 2002) No. 01-1567.
10. Art. 18 Loi No. 2004-800 du 6 Août 2004 Relative à la Bioéthique, modifying Articles L. 613-15 and L. 613-16 of the Code de la Propriété Intellectuelle, *Journal Officiel*, No. 182, 7 August 2004 (also see http://www.legifrance.gouv.fr/WAspad/UnTexteDeJorf?numjo=SANX0100053L).

REFERENCES

Abbott, F.M. (2002), 'The TRIPs Agreement, Access to Medicines and the WTO DOHA Ministerial Conference', *The Journal of World Intellectual Property*, **5**(1), 15–52.

Agreement on Trade-related Aspects of Intellectual Property Rights (1994) (Annex 1C of the Marrakesh Agreement Establishing the World Trade Organization, signed in Marrakesh, Morocco on 15 April 1994), http://www.wto.org.

Andrews, L.B. (2002), 'Genes and Patent Policy: Rethinking Intellectual Property Rights', *Nature Review Genetics*, **3**(10), 803–8.

Beyleveld, Deryck and Robert Brownsword (1997), 'Patenting Human Genes: Legality, Morality and Human Rights', in James William Harris (ed.), *Property Problems. From Genes to Pension Funds*, London: Kluwer, pp. 9–24.

Bobrow, M. and S. Thomas (2001), 'Patents in a Genetic Age. The Present Patent System Risks Becoming a Barrier to Medical Progress', *Nature*, **409**, 15 February.

Bostyn, Sven (2001), *Enabling Biotechnological Inventions in Europe and the United States. A Study of the Patentability of Proteins and DNA Sequences with Special Emphasis on the Disclosure Requirement*, Munich: European Patent Office.

Calabresi, G. and A.D. Melamed (1972), 'Property Rules, Liability Rules and Inalienability: One View of the Cathedral', *Harvard Law Review*, **85**(6), 1089–128.

Chrocziel, P. (1984), 'Zulassungshandlungen mit patentierter Arzneimittelerfindungen durch Zweitanmelder in der Bundesrepublik Deutschland und den USA', *Gewerblicher Rechtsschutz und Urheberrecht, Internationaler Teil* **12**, 735–40.

Clark, J. (2000), 'Patent Pools: A Solution to the Problem of Access in Biotechnology Patents?', White Paper commissioned by Q. Todd Dickinson, the Under Secretary of Commerce for Intellectual Property and Director of the United States Patent and Trademark Office, 5 December, http://www.uspto.gov/web/offices/pac/dapp/opla/patentpool.pdf.

Commission on Intellectual Property Rights (2002), *Integrating Intellectual Property Rights and Development Policy. Final Report*, http://www.iprcommission.org.

Convention on Biological Diversity (CBD) (1992), Rio de Janeiro (Brazil), 5 June 1992 (entered into force 29 December) 1993 (1992), *International Legal Materials*, **31**(818).

Convention on the Grant of European Patents of 5 October 1973 (European Patent Convention [EPC] 1973), as amended by the Decision of the Administrative Council of the European Patent Organisation of 21 December 1978 and of 13 December 1994, http://www.european-patent-office.org.

Cornish, W.R. (1998), 'Experimental Use of Patented Inventions in European Community States', *International Review of Industrial Property and Copyright*, **29**(7), 739–53.

Cottier, T. (1998), 'The Protection of Genetic Resources and Traditional Knowledge: Towards More Specific Rights and Obligations in World Trade Law', *Journal of International Economic Law*, **1**(4), 555–84.

Cottier, T. and M. Panizzon (2004), 'Legal Perspectives on Traditional Knowledge', *Journal of International Economic Law*, **7**(2), 371–87.

Delcorde, A. Claude (1985), 'La Protection des Inventions', in Aimé de Caluwé, *Le droit de la Concurrence*, Volume 2**, Gent: Story, 14.

Ducor, Philippe G. (1998), *Patenting the Recombinant Products of Biotechnology and Other Molecules*, London: Kluwer.

EC Directive 98/44/EC (The Biotechnology Directive) (1998), On the Legal Protection of Biotechnological Inventions, *Official Journal*, L213/13.

Echols, M. (2003), 'Solutions in Terms of Self-regulation or Self-imposed Moral Codes of Business Corporations', in Anton Vedder (ed.), *The WTO and Concerns Regarding Animals and Nature*, Nijmegen: Wolf Legal Productions.

Gold, E.R. (2000), 'Moving the Gene Patent Debate Forward', *Nature Biotechnology*, **18**(12), 1319–21.

Gold, E.R. et al. (2002), 'Needed: Models of Biotechnology Intellectual Property', *Trends in Biotechnology*, **20**(8), 327–9.
Gold, E.R. (2002), 'Biotechnology Patents: Strategies for Meeting Economic and Ethical Concerns', *Nature Genetics*, **30**(4), 359.
Gold, E.R. and T.A. Caulfield (2002), 'The Moral Tollbooth: A Method that Makes Use of the Patent System to Address Ethical Concerns in Biotechnology', *The Lancet*, **359**(9325), 2268–70.
Jacobs, P. and G. Van Overwalle (2001), 'Gene Patents: An Alternative Approach', *European Intellectual Property Review*, **23**(11), 505–6.
Klein, J.I. (1997), 'An Address to the American Intellectual Property Law Association on the Subject of Cross-Licensing and Antitrust Law', http://www.usdoj.gov/atr/public/speeches/1118.pdf, 2 May.
Knoppers, B.M. (1999), 'Status, Sale and Patenting of Human Genetic Material: an International Survey', *Nature Genetics*, **22**(1), 23–26.
Krattiger, A.F. (2004), 'Financing the Bioindustry and Facilitating Biotechnology Transfer', *IP Stragey Today*, No. 8.
Lecrubier, A. (2002), 'Patents and Public Health', *EMBO Reports*, **3**(12), 1120–2.
Leistner, Matthias (2003), 'Traditional Knowledge' in Silke von Lewinski (ed.), *Indigenous Heritage and Intellectual Property*, The Hague, London, New York: Kluwer Law.
Mathély, Paul (1978), *Le Droit Européen des Brevets d'Invention*, Paris: Journal des Notaires et des Advocats.
Merges, Robert P. (2001), 'Institutions for Intellectual Property Transactions: The Case of Patent Pools', in Rochelle Cooper Dreyfuss, Diane Leenheer Zimmerman and Harry First (eds), *Expanding the Boundaries of Intellectual Property*, Oxford: Oxford University Press, pp. 123–65.
Merz, J.F, A.G. Kriss, D.G.B. Leonard and M. Cho (2002), 'Diagnostic Testing Fails the Test', *Nature*, **416**(6872), 577–9.
Moufang, Rainer (1991), *Kommentierung des Europäischen Patentübereinkommen. Artikel 53*, Keulen: Carl Heymanns Verlag, 15. Lieferung.
Mousseron, Jean Marc (1984), *Traîté des Brevets*, Paris: Librairies Techniques.
Murashige, K. (2002), 'Patents and Research. An Uneasy Alliance', in *Public Versus Private Ownership of Scientific Discovery: Legal and Economic Analyses of the Implications of Human Gene Patents*, Special Theme Issue of *Academic Medicine*, **77**(12), 1329–38.
Nelkin, D. (2002), 'Patenting Genes and the Public Interest', *American Journal of Bio-ethics*, **2**(3), 13–15.
Nicol, D. and J. Nielsen (2003), 'Patents and Medical Biotechnology: An Empirical Analysis of Issues Facing the Australian Industry', Centre for Law and Genetics, Occasional Paper No. 6.
Nuffield Council on Bioethics (2002), 'The Ethics of Patenting DNA. A Discussion Paper', www.nuffieldbioethics.org.
Paris Convention for the Protection of Industrial Property (1883), of 20 March, 1883, as last revised at Stockholm on 14 July, 1967, and as amended on 28 September, 1979, www.wipo.int.
Paterson, Gerald (1992), *The European Patent System*, London: Sweet & Maxwell.
Reichman, J.H. (2000), 'Of Green Tulips and Legal Kudzu: Repackaging Rights in Subpatentable Innovation', *Vanderbilt Law Review*, **53**(6), 1743–98 (Symposium Issue), abridged version in Rochelle Cooper Dreyfuss et al. (eds) (2003),

Expanding the Boundaries of Intellectual Property: Innovation Policy for the Knowledge Society, Oxford: Oxford Press.

Resnik, D.B. (2003), 'A Biotechnology Patent Pool: An Idea Whose Time Has Come?', *The Journal of Philosophy, Science & Law*, **3**(1), www.psljournal.com/archives/papers/biotechPatent.cfm.

Saunders, T. (2003), 'Renting Space on the Shoulders of Giants: Madey and the Future of the Experimental Use Exception', *Yale Law Journal*, **113**, 261–8.

Scherer, F.M. (2002), 'The Economics of Human Gene Patents', *Academic Medicine*, **77**(12), 1348–67.

Sevilla, C., J.P. Moatti, C. Julian-Reynier, F. Elsinger, D. Stoppa-Lyonnet, B. Bressac-de-Paillerets and H. Sobol (2002), 'Testing for BRCA1 Mutations: a Cost-effectiveness Analysis', *European Journal of Human Genetics*, **10**(10), 599–606.

Shapiro, Carl (2001), 'Navigating the Patent Thicket: Cross Licenses, Patent Pools and Standard Setting', in Adam B. Jaffe, Josh Lerner and Scott Stern (eds), *Innovation Policy and the Economy, Volume I*, Cambridge: MIT Press, pp. 119–51.

Straus, J. (1993), 'Zur Zulässigkeit klinischer Untersuchungen am Gegenstand abhängiger Verbesserungserfindungen', *Gewerblicher Rechtsschutz und Urheberrecht Internationaler Teil*, **95**(3), 308–19.

Straus, J. (1995), 'Patenting Human Genes in Europe. Past Developments and Prospects for the Future', *International Review of Industrial Property and Copyright Law*, **26**(6), 920–45.

Straus, J. (2003), 'Product Patents on Human DNA Sequences: An Obstacle for Implementing the EU Biotech Directive?', in Scott Kieff (ed.), *Perspectives on Properties of the Human Genome Project*, Amsterdam: Elsevier, pp. 65–78.

Straus, J., H. Holzapfel and M. Lindenmeir (2002), *Genetic Inventions and Patent Law*, Max-Planck-Institüt für ausländisches und internationals Patent-, Urheber- und Wettbewerbsrecht & Bundesminsterium für Bilding und Forschung.

Sung, L. and C. Maisano (2003), 'Piercing the Academic Veil: Disaffecting the Common Law Exception to Patent Infringement Liability and the Future of a Bona Fide Research Use Exception after *Madey* v. *Duke University*', *Journal of Health Care Law and Policy*, **6**, 256–83.

Vandoren, P. (2002), 'Médicaments sans Frontières? Clarification of the Relationship between TRIPS and Public Health resulting from the WTO Doha Ministerial Declaration', *The Journal of World Intellectual Property*, **5**(1), 5–14.

Van Overwalle, Geertrui (1996), *Octrooieerbaarheid van Plantenbiotechnologische Uitvindingen. Patentability of Plant Biotechnological Inventions* (*Substantive Summary*), Brussels: Bruylant.

Van Overwalle, Geertrui (1997), *The Legal Protection of Biotechnological Inventions in Europe and in the United States. Current Framework and Future Developments*, Leuven: University Press.

Van Overwalle, G. (1999), 'Patent Protection for Plants: A Comparison of American and European Approaches', *IDEA: The Journal of Law and Technology*, **39**(2), 143–94.

Van Overwalle, Geertrui (2000a), *Octrooirecht en geneesmiddelen*, Brussels: Bruylant.

Van Overwalle, G. (2000b), 'Klinische Proeven en Volksgezondheid. Naar een Herijking van het Algemeen Belang in het Octrooirecht', *Tijdschrift voor Privaatrecht*, **37**(3), 899–68.

Van Overwalle, G. (2000c), 'The Legal Protection of Biological Material in Belgium', *International Review of Industrial Property and Copyright Law* (IIC), **31**(3), 259–84.

Van Overwalle, G. (2000d), 'Proeven met Geneesmiddelen. De Situatie in België in het Licht van de Recente Rechtspraak en Rechtsleer in Nederland en Duitsland', in Geertrui Van Overwalle (ed.), *Octrooirecht en Geneesmiddelen* (CIR-Reeks 16), Brussels: Bruylant, pp. 65–91.

Van Overwalle, Geertrui (2002), *Study on the Patenting of Inventions Related to Human Stem Cell Research*, European Communities: Luxembourg.

Van Overwalle G. (2003), 'Van Muizen en Mensen. De EU-Biotechnologierichtlijn van 6 juli 1998 en de Omzetting in het Belgisch Octrooirecht', *Rechtskundig Weekblad*, **66**(11), 401–15.

Van Overwalle, G. (2004a), 'Patents in Stem Cell Research in Europe and in the United States', in W. Bender, C. Hauskeller and A. Manzei (eds), *Crossing Borders. Cultural, Religious and Political Differences Concerning Stem Cell Research*, Münster: Agenda Verlag.

Van Overwalle, G. (2004b), 'Van groene muizen en rode oortjes: de EU-Biotechnologierichtlijn en het Belgisch Wetsontwerp van 21 september 2004', *Intellectuele Rechten – Droits Intellectuels* (IRDI), **9**(4), 357–86.

Van Overwalle, Geertrui (2005a), 'Legal and Ethical Aspects of Bio-patenting: Critical Analysis of the EU Biotechnology Directive', in Peter Drahos (ed.), *Death of a Patent System*, Oxon: Law Text Publishing, Chapter 7.

Van Overwalle, Geertrui (2005b), 'Experience on the Patentability of Human Stem Cells and Stem Cells Derived from Them', in *The Ethics of Patenting Human Genes and Stem Cells*, Copenhagen: Council of Ethics.

Van Overwalle, G. (2005c), 'Protecting and Sharing Biodiversity and Traditional Knowledge: Holder and User Tools', *Ecological Economics*, **53**, 585–608.

Walsh, John P., Ashish Arora and Wesley M. Cohen (2003), 'Effects of Research Tool Patents and Licensing on Biomedical Innovation', in Wesley M. Cohen (ed.), *Patents in the Knowledge-based Economy*, Washington: National Academic Press, pp. 285–340.

Wolfrum, Rüdiger and Peter-Tobias Stoll (2002), *Die Gewährleistung freier Forschung an und mit Genen und das Interesse an der wirtschaftlichen Nutzung ihrer Ergebnisse*, Frankfurt: Verlagsgruppe Peter Lang.

Index

not met, ethical approval will not be granted, or if it has been granted, ongoing research should be stopped when the problem becomes known. Moreover, civil remedies would be available for aggrieved research subjects as discussed above, and indeed this was precisely the result in that applied in Moore. But as we have seen, where there is no connection between the regulation and patent systems such practices leave a patent on related research products unaffected – a patent that facilitates commercial exploitation of an invention that has been developed through unacceptable practices and unauthorised instrumentalisation of research subjects. This suggests that the claim of redundancy may not be well founded in these circumstances. Rather, regulating requirements of informed consent to the validity of a patent could, in fact, provide an additional level of protection achieved through the threat of patent denial or revocation lest the requisite consent has not been obtained. The patent system, whose so often role is driven by incentive. If this is true, cannot the threat of patent denial or revocation act as a disincentive to unethical conduct? This would be a forceful message about the unacceptability of certain practices and it might have a positive influence on such practices for the future. Even if the patentee is not the researcher or person who obtains consent, it is in each party's interests to ensure that consent, assuming a professional/commercial relationship between them. And although the mere denial of a patent does not represent any direct remedy for those research participants who have been disrespected, it serves both as a sanction against those who have disrespected them and ensures that the key object of concern – the economic advantages represented by a patent – is thwarted.

A more worrying aspect, however, is that patent denial or revocation cannot prevent use of the material obtained from research participants. Does this, then, fundamentally undermine the approach? Arguably not. Rather, it highlights the limitations of the patent system in *regulating* science and scientific conduct. It should, instead, re-enforce the message that while overlap exists between the systems, the patent system, for its part, has a restricted role to play. This does not mean, however, that it cannot bring something valuable to the overall approbation of the research trajectory, so long as its supporting role is recognised and additionality is achieved. The contribution is not in the realm of direct regulation but in *incentivisation*, or perhaps more accurately, *disincentivisation*.

8.3 Achieving Coherence

The patent system does not exist in a moral vacuum. This is reflected in the wording of Article 6 of the Biotechnology Directive, which states:

inventions shall be considered unpatentable where their commercial exploitation would be contrary to ordre public or morality; however, exploitation shall not be deemed to be so contrary merely because it is prohibited by law or regulation' (EC Directive 98/44 1998). The morality provisions can act as a mechanism to register opposition to the prospect of commercial exploitation of particular inventions. Exploitation is precisely the concern at the heart of consent-based objections to commercialization practices in research: 'had we been told we would (or might) have refused' – and the commercial exploitation of inventions derived (in part) through practices that disrespect research participants is therefore also highly questionable. According preference to those inventions compounds the affront. To the extent that we accept there is an overarching obligation to uphold the fundamental principles of respect for human dignity and the right to [illegible] the fruits of unethical research practices [illegible]. This is an issue of seeking congruence between regulatory and patent regimes. Where the regulatory system has already mandated consent, and where a consent provision also appears in the patent law, congruence would suggest that each system should reflect and re-enforce this value preference in a manner that is true to the essentials of that preference. Thus, if consent is about ensuring that individuals can exercise their self-determination within a social framework that respects them and values their choices, one would expect disrespectful conduct to be sanctioned, or at least not supported by the state in contradictory fashion as between different segments of that framework. To do otherwise exposes these state-controlled systems to charges of inconsistency and, ultimately, challenges of legitimacy. While this is not to say that different factors and considerations do not apply in different contexts and circumstances – for the patent system is not the regulatory system nor vice versa – the clear and strong commitment in Europe to the well-accepted fundamentals of human dignity and integrity across both of these systems points to a need for the congruence spoken of here. At the very least, it places the onus on Member States to justify examples of incongruence, something that has not happened to date to any acceptable degree.

Finally, it is also noteworthy that the second part of Article 6(1) returns us to considerations of additionality in that it provides a further reason to seek sanctions from within the patent system itself. For, while it could be argued that the regulatory system is perfectly well equipped to impose commercial sanctions on unethical researchers in the guise of licence refusal, effectively ruling out a domestic market, the ongoing availability of protection would remain an incentive to produce domestically under patent for exportation and exploitation abroad as the EPO itself